AF393322

Technische Universität München
TUM School of Engineering and Design

Energy Systems Optimization Considering the Uncertainty of Future Developments

Wolf Gereon Wedel, M. Sc.

Vollständiger Abdruck der von der TUM School of Engineering and Design
der Technischen Universität München zur Erlangung des
Doktors der Ingenieurswissenschaften
(Dr.-Ing.)
genehmigten Dissertation

Vorsitz:	Prof. Dr. phil. Klaus Bengler
Prüfer der Dissertation:	1. Prof. Dr.-Ing. Hartmut Spliethoff
	2. Prof. Dr. rer. nat. Thomas Hamacher

Die Dissertation wurde am 28.06.2023 bei der Technischen Universität München eingereicht und durch die TUM School of Engineering and Design am 21.12.2023 angenommen.

Bibliografische Information der Deutschen Nationalbibliothek:

Die Deutsche Nationalbibliothek verzeichnet diese Publikation in der Deutschen Nationalbibliografie; detaillierte bibliografische Daten sind im Internet über **dnb.dnb.de** abrufbar.

Die automatisierte Analyse des Werkes, um daraus Informationen insbesondere über Muster, Trends und Korrelationen gemäß §44b UrhG („Text und Data Mining") zu gewinnen, ist untersagt.

Herstellung und Verlag: BoD – Books on Demand, Norderstedt

1. Auflage 2024

ISBN: 978-3-7597-0293-7

Preface

Die vorliegende Dissertation entstand im Rahmen meiner Tätigkeit als wissenschaftlicher Angestellter am Lehrstuhl für Energiesysteme der Technischen Universität München. Besonders bedanken möchte ich mich bei Herrn Prof. Spliethoff für die Möglichkeit und das Vertrauen an seinem Lehrstuhl zu forschen und diese Arbeit anzufertigen, sowie den Gestaltungsspielraum den ich genießen konnte. Außerdem möchte ich mich für seine beeindruckende Fähigkeit fachlich und menschlich hervorragende Mitarbeiterinnen und Mitarbeiter auszusuchen bedanken, ich schätze mich glücklich eine ganze Reihe von diesen zu meinen Freunden zählen zu dürfen. Während meiner Zeit am Lehrstuhl haben viele Mitarbeiter*innen dort angefangen und aufgehört. Jede und jeder von euch hat dazu beigetragen diese Zeit zu etwas ganz Besonderem zu machen. Dafür vielen vielen Dank!

Ganz besonders möchte ich mich bei Sebastian M. für die Unterstützung bezüglich der Optimierung, sowie die Automatisierung der Berechnung des COP bedanken. Bei Hans bedanke ich mich für seine Fachkenntnisse und Erläuterungen zur Jensen-Ungleichung. Andi H. möchte ich für seine Arbeit, erst als Student und nachher als Kollege danken. Dafür jeden Arbeitstag zu einem Tag unter Freunden gemacht zu haben möchte ich der 15 Uhr Kaffeerunde bestehend aus Moritz G., Felix, Julia, Michi A., Babsi, Jell und Manu., sowie dem Traderclub bestehend aus Vinni, Moritz, Philipp und Sebastian M. danken (Laura, wenn es dich damals schon gegeben hätte, würdest du hier bestimmt auch auftauchen). Felix außerdem vielen Dank für die vielen tollen Läufe und die Energie immer wieder alle zur Teilnahme an der Laufrunde zu motivieren. Lynn, Du warst die beste Bürokollegin, die ich mir vorstellen kann, danke für die vielen angenehmen und produktiven Stunden. Meinem Projektkollegen Benedikt möchte ich für die effektive Zusammenarbeit, den Versuch die Welt auch im Kleinen immer ein bisschen besser zu machen und das Korrekturlesen dieser Arbeit danken. He-Man, Sebastian E. und Claudia möchte ich für eine unvergessliche Konferenz sowie den technischen bis philosophischen Austausch danken. Dem He-Man soll außerdem hier für unvergleichliche Freude auf der Tanzfläche erwähnt werden, danke für die vielen schönen Zeiten! Moritz B., vielen Dank für deinen Humor, Blobby-Volley und die guten Unterhaltungen. Richi, PJ und Tobias G. vielen Dank dafür ein durchgehend positives Beispiel körperlicher Fitness gepaart mit einem scharfen Verstand und vielen guten Gesprächen gewesen zu sein. Michi H. danke ich für seine Ruhe und Gelassenheit, sowie ihm und jedem sowie jeder, der oder die Teilnehmer der LES-Band war, für die vielen glücklichen Stunden durch eure Musik. Flo und Peter vielen Dank für fußballerische und menschliche Meisterleistungen. Für glückliche Momente und gute Gespräche möchte ich mich außerdem bei Kristina, Gesa, Tobias N., Andi S., Roberto, Matthäus, Thorben, Clemens und Härzschel bedanken. Bestimmt habe ich jemanden vergessen, danke auch an Dich, falls Du dazu gehörst! Ein ganz besonderer Dank gilt auch den Kollegen*innen, die am Ende meiner Zeit und nach mir kamen und mich bei jedem Besuch am Lehrstuhl wärmstens willkommen heißen. Auch meinen Freunden außerhalb des Lehrstuhls möchte ich danken, dass sie während dieser ereignisreichen Zeit immer für Austausch und Unterstützung zur Verfügung standen.

Vielen Dank an meine Eltern für Vertrauen und Freiheit. Außerdem danke ich meinen Schwestern, einfach dafür, dass ihr da seid, ihr seid die besten. Grazie anche a Tiziana e Bruno per l'aiuto e per permetterci di goderci al meglio la bella vita e la nostra famiglia. Eleonora und Olivia, ihr zeigt mir jeden Tag was wirklich wichtig ist im Leben, danke dafür.

Kurzfassung

In Anbetracht des fortschreitenden menschenverursachten Klimawandels und der Rolle von Energie bei der Sicherung eines hohen Lebensstandards für die Weltbevölkerung, wurde und wird die Energiesystemoptimierung genutzt um Erkenntnisse zu unterschiedlichen, möglichen Strukturen des Energiesystems zu gewinnen. Ein Hauptaspekt der aktuellen Forschung ist es kosteneffiziente Energiesysteme zu identifizieren, die eine Senkung der Treibhausgasemissionen erlauben. In diesem Kontext wird im Rahmen dieser Arbeit untersucht wie sich die Unsicherheit in Bezug auf zukünftige Technologiekosten auf die Ergebnisse von Energiesystemoptimierungen auswirkt. Erreicht wird dies mittels stochastischen Optimierungen mit dem Ziel der Kostenminimierung unter Berücksichtigung von Wahrscheinlichkeitsverteilungen. Diese repräsentieren erwartete, zukünftige Technologiekosten und deren zugrundeliegende Unsicherheit. Basierend auf theoretischen Überlegungen und einem vereinfachten Beispielenergiesystem wird gezeigt, dass die jensensche Ungleichung dazu führt, dass Szenariooptimierungen, die nur den Mittelwert der zukünftigen, erwarteten Kostenverteilung berücksichtigen, die notwendigen optimalen Systemkosten systematisch überschätzen.

Ein Energiesystemoptimierungsmodell für Deutschland, welches Wärme-, Elektrizitäts- und Verkehrssektor berücksichtigt, wird formuliert und stochastische Optimierung darauf angewendet. Die Ergebnisse werden mit den Szenarioergebnissen verglichen. Es wird untersucht in wie weit ein Metamodell basierend auf Faktoreffekten oder die Verwendung von weniger Optimierungen für die stochastische Optimierung eine Verringerung des Rechenaufwands ermöglichen. Die Ergebnisse bestätigen ein Überschätzen der notwendigen Systemkosten, wenn der Szenarioansatz basierend auf dem Mittelwert der erwarteten Kosten genutzt wird. Im Fall einer vorgegebenen minimalen Reduktion der CO_2-Emissionen um 80 % gegenüber 1990 werden die notwendigen Kosten um ca. 3.5 % und im Falle eines vollständig erneuerbaren Systems um 0.4 % überschätzt. Die stochastische Optimierung liefert den Interquartilsabstand, der genutzt werden kann um die Unsicherheit des Ergebnisses zu charakterisieren. Dieser liegt für das komplett erneuerbare System zum Beispiel bei 13.2 € MWh^{-1} bzw. bei 27.3 % des Mittelwerts.

Von den untersuchten Möglichkeiten den Rechenaufwand zu verringern zeigt die Nutzung von 30 bis 60 Optimierungen anstatt der 500 Optimierungen, die als Benchmark genutzt werden, vergleichbare Ergebnisse. Die Nutzung der betrachteten Metamodelle hat, abgesehen von der Möglichkeit extreme Ergebnisse vorherzugsagen, im Vergleich dazu keine Vorteile.

Für einige Ergebnisgrößen, insbesondere in den untersuchten nicht erneuerbaren Fällen, weicht der Erwartungswert, der aus der stochastischen Optimierung resultiert, stark von dem Ergebnis der Szenariooptimierung ab. Ein Beispiel dafür sind die CO_2-Emissionen, von denen in einem kostenoptimalen System nach der Szenariooptimierung nur ca. 18 % des Erwartungswerts der stochastischen Optimierung emittiert würden. Ähnliche Abweichungen können für andere Ergebnisgrößen identifiziert werden auch wenn die meisten Erwartungswerte durch den Szenarioansatz relativ genau approximiert werden.

Die Anwendung von Algorithmen zur Klassifizierung der Ergebnisse zeigt sich nützlich um die Vielzahl der unterschiedlichen aus der stochastischen Optimierung resultierenden Systeme auszuwerten und z. B. zugrundeliegende Systemstrukturen zu identifizieren. Die inhärente Robustheit der stochastischen Optimierung gegenüber kleinen Änderungen der Eingangsgrößen erhöht das Vertrauen in die Ergebnisse. Dies ist besonders relevant, da der Einfluss von geänderten Einflussgrößen bei Szenariooptimierungen, die die meisten veröffentlichten Studien darstellen, aufgrund der Komplexität der Optimierung nicht a-priori bestimmbar ist.

Abstract

In the light of anthropogenic climate change and the importance of energy to ensure high standards of living, energy system optimization has been and is used in order to gather knowledge regarding different possible energy system layouts. One main driver in recent years has been determining possibilities to cost efficiently mitigate greenhouse gas emissions. In this context this work investigates the influence future uncertainties regarding technology costs have on optimization results. This is achieved by performing energy system optimization with the optimization objective of reducing system cost using stochastic optimization with underlying probability distributions to capture expected future cost and its uncertainty. Based on theoretical considerations and a minimal example energy system, it is shown that Jensen's inequality leads to an overestimation of necessary, optimal system cost when a scenario optimization taking into account only the expected technology cost means is used.

Stochastic optimization is used on a herein constructed model of the German energy system that comprises electricity, heating and transport sector. Stochastic optimization results are compared to the corresponding scenario results based on the cost distributions means. The derivation of a factor effect based meta model as well as the use of fewer optimizations in stochastic analysis is investigated as means to reduce the computational effort of the proposed methodology. The results confirm the overestimation of necessary cost achieved by scenario optimization, in the complex example by about 3.5 % at the boundary condition of at least 80 % emission reduction compared to 1990 and 0.4 % if the system is completely renewable. Stochastic optimization also yields the inter quartile range which can be used to characterize uncertainty. In the case of a completely renewable energy system the inter quartile range is 13.2 € MWh^{-1} respectively 27.3 % of the mean.

From the investigated possibilities to reduce computational demand the use of 30 to 60 optimizations in the stochastic case yields similar results compared to the use of 500 optimizations, which serve as benchmark. The use of the proposed meta models does not yield significant advantages apart from the possibility to predict extreme results, which do not show up in the reduced case with 30 to 60 underlying optimizations.

For some result parameters, especially in the not completely renewable cases, the expected value from stochastic optimization differs greatly from that achieved by scenario optimizations. One example are expected carbon emissions at an emission limit of 20 % of 1990 emissions. In this case the scenario optimization yields only about 18 % of the CO_2 emissions that result as the the mean of stochastic optimization. For other parameters similar differences are revealed while most parameters are represented well by the scenario results.

Clustering is shown to be useful to manage the plethora of different results from stochastic optimization as it allows to identify underlying system layouts. Stochastic optimization with probability distributions are inherently robust as small changes to the distributions effect the outcome only little. This allows strengthening trust in results as usually readers of energy system studies do not agree with the authors on cost assumptions and the implications of input parameter changes are not predictable due to the complexity of the topic.

Contents

Contents

Contents

X

List of Figures

List of Tables

Nomenclature

Abbreviations

Chapter 1

Introduction and Motivation

Energy is important. This is expressed in the use of energy related vocabulary and expressions in everyday life relating to our emotional and physical well being. The development and increase of energy use has historically been closely linked with the improvements in standard of living and economic development [1]. According to MacKay the use of 1 kWh d^{-1} is comparable to the availability of a human servant [2]. With a final energy consumption of about 2.5 PWh and a population of 82.9 million in 2018 [3] this equals about 85 servants for each person that lives in Germany, not including the energy expense of imported goods. Even though in the last decades a decoupling of energy consumption and the development of economic indicators as e.g. the gross domestic product has been observed [4], there is no doubt that availability of energy is crucial for achieving a high standard of living for the world's population.

Research in energy systems in the past has been motivated mainly by the need to ensure security of supply. While this is still a relevant topic, in recent years the public consensus based on scientific evidence that anthropogenic climate change is taking place, has caused the mitigation of greenhouse gas emissions to become one of the most important motivations to investigate energy systems as well as to propose and plan possible futures of energy supply.

1.1 Energy System Optimization

Energy system optimization, i.e. a matching of energy supply and demand in an mathematical optimization model, is a useful tool for energy system planning and widely used in government and industry. Many different aspects spanning local to global energy supply, sometimes also including non energy sectors which interact with the energy sector, have been investigated using energy system optimization models.

1.1.1 Uncertainty in Energy System Optimization

Despite or because of the high value energy delivers, the goal of energy system optimization is usually to propose an energy system that supplies energy at the lowest cost, as a measure for efficiency. One of the great advantages of energy system optimization is the possibility to internalize currently externalized costs such as environmental damages from air pollution or the global damages caused by climate change. For these costs the German environmental agency recommends the the use of climate damage costs of 180 to 730 € $t_{CO_2}^{-1}$ depending on the time of emission and how much of the cost inflicted on future generations is taken into account [5].

As the goal of energy system optimization is frequently to give design guidelines for a cost efficient future energy system, there is the necessity to assume values of future properties of the energy system. These are especially technology costs and conversion efficiencies. Any assumed value in this context is, without doubt, heavily uncertain, especially concerning costs. There are many methodologies to systematically derive future costs, and yet there is scientific evidence, that any estimate is always only a guess and for more than a couple of years in the future the prediction quality is basically that of chance [6].

1.1.2 Opportunities and Challenges of Stochastic Optimization of Energy Systems

The term stochastic optimization is not always used with the same meaning. Herein solving an optimization problem many times in order to achieve distributions of the output parameters is intended. This procedure allows for a discretization of the uncertainty regarding input parameters in the form of probability distributions. Therefore stochastic optimization enables both, to evaluate the effect of uncertainty on probable optimal future energy systems and to gain more information from an optimization model compared to a scenario analysis with a set of different scenarios.

Regarding the use of stochastic optimization there are some challenges which might be the cause for the not yet wide spread use of this methodology. These challenges include the communication and interpretation of results and assumptions. In the case of stochastic optimization all these are not single figures but distributions which makes their elaboration much more complex. Another drawback to stochastic optimization is the higher computational effort compared to a scenario analysis due to the necessity to run the optimization many times with different input parameter combinations. Generally models are becoming more complex and detailed, which leads, in the perception of the author, to little or no improvement regarding the computational time required to run an optimization as advances in computing are compensated.

In order to highlight the advantages and contribute to the possibility for wide spread use of stochastic optimization in energy systems some investigations are carried out. These include an investigation of systematic errors when not considering uncertainty as well as the effect and magnitude of the amount of uncertainty. This is expressed by the number of uncertain parameters and the probability distribution widths. Additional to these investigations on a simplified energy system model, a more complex and, in terms of comparability to state of the art energy models, more relevant model is constructed. It models the German energy system and is used for stochastic and scenario optimization. The results are compared and evaluated regarding the information gain. In order to address the before mentioned challenges of interpretability and communication regarding stochastic optimization, factor effect methodology and clustering algorithms are applied. These are used to investigate possible insights regarding the dependencies of optimal configurations from the uncertain parameters and to reduce complexity by introducing clusters with similar properties regarding the energy system structure. This allows for easier communication of energy system optimization results and additional insights, as the structure of the energy system itself, which for many decision makers is one of the important information from energy system optimization, is the criteria for clustering. To enable the use of stochastic optimization also for complex models which do now allow for many model runs, meta model derivation based on factor effect methodology is investigated alongside the use of distributions consisting of fewer optimizations. The comparison reveals comparable information gain for both approaches, with higher accuracy when stochastic optimization with smaller distributions is used compared to meta models based on the same number of optimizations. It is demonstrated that the use of less optimizations enables the computational effort to be reduced significantly while retaining many of the advantages of stochastic optimization.

Chapter 2

State of Knowledge

This chapter is intended to give insights into the development and current state of research regarding energy system optimization and the consideration of uncertainty. When using the term energy system optimization herein the focus is not on the technology scale as e.g. engines or gas turbines including thermodynamic modeling etc., but instead the energy system itself, providing energy in a spatial and temporal context to consumers when it is needed. In the first section of this chapter a definition of energy system planning is given. Furthermore the historic development of energy system optimization and the different aims with which research has been and is performed as well as the different model types and methods that are used are summarized. The research focused on uncertainties in energy systems, its origins, quantification and how it is accounted for in energy system optimization is part of the second section of this chapter.

2.1 Energy System Optimization (ESO)

Historically the development of ESO was catalyzed by two trends. On the one hand this was the growing complexity of energy systems including a mix of generation technologies such as hydro, pumped hydro and thermal power plants as well as more complex transmission grids [7]. The other important trend was the development of the necessary mathematical tools [7] which has ultimately led to ESO as we understand it today. Pumped hydro power plants e.g. introduce storage into the power system, the optimal operation of which is not trivial. An important milestone in the development of ESO was the definition of "System Planning" by a subcommittee of the Edison Electric Institute in 1953 that was disseminated by the Edison Electric Institute and the American Institute of Electrical Engineers [8].

"System planning is the preparation of a rational program for the development of an electric power system, so that it can evolve in an orderly and economic manner. It includes forecasting and analyzing loads, rationalizing standards of service, anticipating trends in equipment design and coordinating the various elements of the system into a well-designed whole; it is particularly concerned with plans for changes and additions to generation, transmission, substations and distribution facilities. It is not concerned with the problems of day to day operation or design except to the extent that these problems effect future system development. Briefly, electric system planning is the process of determining when, what facilities should be provided where in order to assure adequate electric service at minimum average annual cost to the community." [8]

Contrary to the definition from the Edison Electric Institute, in the early stages the upcoming techniques have been used especially for power plant scheduling and plant operation. Kirchmayer [7] gives an overview over the scientific developments from 1942 to 1950. These include economic loading, reduction of transmission losses, the representation of incremental fuel costs and the necessary mathematical methods for optimized scheduling, automatic operation of networks and dispatching. All these advances have been made with the goal to lower the cost of providing electricity [7]. Kirchmayer himself contributed widely to the field of operation planning, the publications [9–11], co-authored by him, all regarding hydroelectric plants, their operation or interaction with other power plants, have each been cited over 20 times with the most recent citations in 2018 and 2019, showing the continued relevance of the approaches. The mathematical basis for ESO as we know them today, originating from the years 1949 to 1963, have been presented in 1963 by Dantzig [12]. This includes the introduction of the term linear programming for linear inequation theory with the goal of minimization. He states that what linear equation theory had been for the natural sciences, linear inequation theory, i.e. linear programming, has become for decision problems [12].

In 1974 Finon describes a mathematical, linear programming model of the French energy sector [13]. The optimization model spans the years 1976 to 2020, divided in various sub periods. It includes energy conversion processes, distribution and storage, as well as different types of consumption including transport, residential and industry sector. The demand side is simplified via a representation of three demand configurations derived from time series analysis. The three load cases are peak load, i.e. the 120 hours with the highest load, base load and the "critical period", which Finon identified as most influencing for the plant choice and which corresponds to a 800 to 1200 hour per year period with high load, inferior to that of peak load. With this representation in order to allow for an optimization of the installed capacity of energy infrastructure and conversion technology, Finon performs system planning according to the definition by the Edison Electric Institute [8]. The basis of the model is that energy supply and demand must be balanced during each period. Many current ESO models work similar to Finon's model, their underlying structure is depicted in fig. 2.1. This structure is adapted to the different aims with which models have been developed. This combination of system planning with optimization is called ESO. As Finon's model contains the simplification of only three load conditions, it is not suited to optimize volatile electricity production e.g. by photovoltaic or wind power plants. As these have become more important during the last decades, many models with a higher resolution in time have been developed.

A basis for the wide spread use of ESO is the ESO framework MARKAL, published by the International Energy Agency, founded in 1974. Even though other frameworks exist MARKAL is one of the most influential. The first

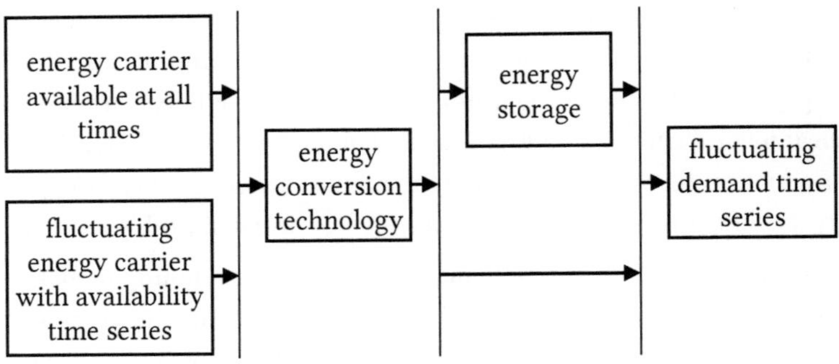

Figure 2.1: The underlying structure of ESO models

efforts to develop an ESO model at the IEA began in 1976 with its *Energy Technology Systems Analysis Program.* The linear programming framework MARKAL was first presented at the Energy Systems Analysis International Conference in 1979 by Fishbone [14] and later published in the International Journal of Energy Research [15]. In MARKAL, optimization spans several decades and constraints are implemented, including that the demand of an energy form must be met by production or by supply from storage in each time step. Additional constraints can be introduced, the MARKAL framework e.g. contains constraints regarding the maximum capacity of district heating networks and there is the possibility to implement a constraint on growth of annual investment costs [14]. A typical constraint in more recent studies is the definition of maximal carbon dioxide emissions during the optimized period. In contrast to Finon, six typical electrical demands are identified by time series analysis that takes into account yearly and daily fluctuations. The availability of a technology during peak periods is considered with the use of availability factors. [15]

2.1.1 Aims of Energy System Optimization

ESO has been, and is used for different purposes ranging from the design of energy systems on the level of neighborhoods [16] to determining political pathways for energy system development on an international level [17]. MARKAL has been one of the first ESO model-frameworks with wide spread use. Extensions of it e.g. the Global Multi-regional MARKAL model [18] and TIMES [19] are still relevant today. Therefore it is suited for exemplary investigation of the evolution of ESO use, here with a focus on national or state level analysis. An analysis of an excerpt of the 355 publications listed in Scopus at the time of writing that cite the original MARKAL publication [15] enables us to determine a trend over time of the main aims with which ESO have been used. In the publication describing MARKAL [15], the goals with which the model has been created are named:

1. Comparing current and future energy technologies and recourses regarding how they satisfy expected future energy demands.

2. Evaluate future development of implementation and costs of technologies as well as resources and the change of current resource utilization (e.g. petroleum).

3. Gain knowledge about the sensitivity of an optimal future energy system regarding fuel costs and technology cost development.

4. Assessment of the effects efficiency improvements have on the energy system.

These objectives are in general the aims of the specific publications discussed hereinafter, as these are specifications of these goals. An additional objective that is frequently mentioned in the publications compared is the evaluation of political strategies or the definition of such, using the insights gained with the model. A more recent account of the various uses the MARKAL model has been put to is given in [20]. Additional to the above the following applications are reported:

5. The identification of investment strategies based on least-cost energy systems.

6. Finding solutions to comply with environmental restrictions in a cost efficient manner.

7. Gain insight into effective prioritization of research and development regarding emerging technologies.

8. Definition of greenhouse gas emission reduction goals and evaluation of emission trading considering the context of international cooperation.

Two relatively early publications that cite the original MARKAL publication [15], one by Cofala [21] in 1985 and one by Luthra and Fuller [22] in 1990, mirror the broad variety of ESO applications. Both articles present different aims that are achieved with similar measures. The first aim is governmental planning in order to allocate resources in an economically efficient manner [21]. The other aim is private and public research with the goal to scrutinize policy decisions[22]. The resason for this is given by [22] as high imports of fossil fuels, about 75 percent of primary energy in the considered region at that time, combined with an "uncertain oil future". Cofala uses a linear optimization model of Poland, which is structurally based on MARKAL [15] and the similarly influential MESSAGE model [23]. MESSAGE is an example for integrated models that include other sectors besides energy, published and used by the International Institute for Applied Systems Analysis. According to Cofala, special care is taken to assure consistent scenarios between all the considered parameters. Luthra and Fuller use existing models with the goal to combine these in the MARKAL framework from [15]. They use the framework WATEMS [24] to simulate the future development of Ontario's energy system and to estimate the social cost of policy constraints [22]. Both investigations include sub models for economic development and various economic sectors in order to capture the interdependencies between energy, economy and demography. While Cofala [21] uses 5-year periods for the optimization and development forecast over a period of 20-25 years until 2005, Luthra and Fuller [22] base their analysis on an annual optimization and comparison of the years 1984 and 2021. By using technology and demand forecasting and including preliminary steps in the energy supply chain, such as refinery processes, Luthra und Fuller investigate the boundary conditions for a fuel switch to biofuels based on sensitivity analysis. In order to identify desirable changes regarding economics and air-pollution the ESO results are compared to the actual 1990 energy system. Their overall goal is to test the influences of policy decisions, in this case not installing new nuclear power plants, higher insulation standards and renewable heating for homes, with regard to the economics and the structure of the resulting energy supply system and energy supply itself.

Research from Mattsson and Weene [25] published in 1997 has an altogether different goal from the quantification of possible outcomes of specific policy decisions for a country or region persued by [21, 22]. They identify the need to evaluate the future value of technologies and their impact on a system in order to decide on research and development as well as energy strategies, e.g. in the light of global warming [25]. Additionally they state that it is necessary to take into account future developments and the mutual influences investment decisions and technology costs have on one another [25]. For the investigation of these influences they formulate a model that considers learning, respectively experience curves based on the insights from Bruce Henderson [26], founder of the Boston Consulting Group. The improvement compared to previous research is that the experience curves are considered already in the model formulation. This results in a non-convex model, which takes into account changing technology costs according to experience curves while optimizing installed capacities to reach lowest overall cost. In order for this model formulation to be able to describe global developments the whole world, divided into four regions is considered. The optimization is carried out for the period from 1990 to 2050, such that the present value of total cost is minimized. The aim of the research is uncovering the dynamics of technology engineering costs in order to allow for an analysis of energy policy, and to generate scenarios for more detailed models. This approach allows avoiding technology lock in and thus, according to the authors, enables to mitigate climate change as renewable technologies are implemented now in order to decrease costs in the future and achieve a lower overall cost.

A study with the purpose to analyze scenarios resulting from policy decisions and their effect on a countries energy system was performed by Andersson and Hådén in 1997 [27]. This is an example of research that has been strongly influenced by ESO, pursuing similar goals, even if no energy system model is used in the research itself. The authors are motivated by legislation that has been passed previous to their research. According to [27] that legislation comprises a complex set of objectives that are difficult, or as they state "impossible to fulfil" in the future. The specific policy decisions include a nuclear phase out resulting from a referendum held in 1980 and CO_2 emission freeze on 1990 emission levels. Additional to this also the decision against development of hydro power in, until that point, unharnessed rivers in combination with an economic and secure electricity supply are cited. Under this complex set of seemingly contradictory goals, the authors make use of an energy market model that combines electricity and heating sector, including also light and heavy oil. Differing from the previously mentioned [15, 23], which are bottom-up models, it is based on market structures. The differentiation between model types is layed out on page 17. They take into account demand elasticities depending on the price for energy and structure their model according to the structure of Sweden's heating market. The only exogenous costs are the price of heavy and light oil; the model determines all other prices as equilibrium prices under the assumption of perfect competition. To take into account the varying load conditions, three different load cases are assumed, similar to the assumptions taken by Finon [13].

Evidence of the importance of the MARKAL model is the number of extensions to the original model that have been developed. One of these is the Global MARKAL Model (GMM) developed by Barreto [28]. An application of this model with the goal to evaluate different policies regarding their environmental impact is presented by Rafaj et al. [29]. The three policy measures evaluated in order to limit CO_2 emissions or air pollution are Cap-and-Trade of carbon emissions via a reduction target for the global energy system, imposing a minimum share of renewables in the electricity production portfolio and internalizing cost of air pollution associated with power generation. An additional goal is the identification of synergies and tradeoffs between the different policy scenarios. The GMM is a bottom up model of the energy sector, containing endogenized technology learning and a detailed representation of energy technologies. Electricity as well as heating and transport sector are included. The authors are motivated by the need for a long term strategy in order to transform the energy system from its current status to a more sustainable version. The aim of this investigation of possible effects of policy measures is to support the political decision-making process.

Another example for the use of a MARKAL model is [30]. Shearer et al. use the US-MARKAL model to explore the effects of natural gas supply on renewable energy use and CO_2 emissions in the United States to check if the promoted solution for decarbonizing the energy sector by means of natural gas from fracking effectively leads to an emission reduction. Their motivation is that natural gas is many times promoted as a low cost solution for carbon emission reduction. Yet they cite an increasing number of studies that suggest, even though carbon emissions may be reduced in the short term, significant long-term reductions without climate policy are not possible. To take into account the special US situation, leakage during natural gas production with hydraulic fracturing technology is considered regarding greenhouse gas emission impact. In order to adequately capture future developments expert elicitation methods for natural gas supply curves are used. The sensitivity of the US energy system to different gas supply scenarios is considered in combination with three contrasting climate policies. Those are namely the absence of a specific climate policy, carbon tax introduction and a strict carbon cap combined with a federal renewable program. Energy demand elasticities are not taken into account.

The same overarching goal, the investigation of measures for carbon emission reduction, is the motivation of Sandvall et al. [31]. Specifically they see global urbanization asking for throughout assessment of different heating and cooling options in newly build urban areas, as heating and cooling in 2014 represented 36 % of greenhouse gas emissions in the EU and more and more people globally are moving to live in cities [31]. The climate impact of low-energy building areas is evaluated depending on heat supply, which can be realized either by connection to neighboring district heating systems or by decentralized supply options. The model used for this research is the TIMES-UH model [32], which is based on the MARKAL extension TIMES [19] and optimizes only the heating sector. Electricity, transport and residential sector values have to be supplied as input data. The optimization objective is cost minimization, which is performed for two regions defined by their heat supply options. The simulation is performed as multi-period optimization with ten time steps between 2014 and 2052. The approach to daily heating load curves is similar to that described in [15] for electricity, instead of six typical demand situations eight time slices are used.

Considering the publications discussed above, a trend in the motivation behind the use of energy system modeling is discernible even in this limited excerpt of the literature. The earlier publications aim at reducing fossil fuel dependency as an answer to the oil crises, which lead to the first development of such models [15], whereas more recent research efforts are directed towards mitigating climate change and to define policies for energy systems with reduced greenhouse gas emissions [33].

2.1.2 Methods and Models

In this sub chapter an overview on the variety on energy system models will be given, with no claim of completeness as it is a complex, evolving field and numerous models and applications exist. This view was already published in 1999 by Van Beeck [34], together with an overview over the various attempts made to characterize and categorize models. Since then more models have evolved, many of which are multipurpose and can be used in various contexts and with different aims. Van Beeck states rising computational capabilities as one of the drivers of model development [34]. This is reflected by looking at the change of model structure over time with one example being the use of typical days or typical load periods to reduce necessary temporal resolution especially in the earlier models, e.g. [15, 23]. Now even multi-decade models can contain at least base years in detailed hourly resolution for several types of demand, e.g. electricity, heat and cold, and still be solved with the assumption of perfect foresight on personal computers within a reasonable time. One example is the intertemporal urbs model developed during the Clean Tech Campus project from 2016 to 2019 at the Chair of Sustainable and Renewable Energy Systems and the Chair of Energy Systems of Technical University of Munich [35, 36]. In order to structure the plethora of different approaches and implementations the classification approaches summarized by [34] will be followed by classifying the models according to the following aspects:

1. Purposes of energy models

2. Model structure regarding internal and external assumptions

3. Top-Down and Bottom-Up

4. Methodology

5. Mathematical Approach

6. Geographical coverage

7. Sectoral coverage

8. Time horizon

9. Data requirements

As mentioned by [34] these are not independent of one-another, often one aspect leads to certain tendencies in others, nor are these the only ways to classify a model. Yet the classification covers the most important differences and suffices for achieving an overview of the differences between different model architectures. Structure and content of this subchapter are based on [34] with some additions.

Purposes of energy models

Based on the analysis performed by [34, 37] three general purposes and four specific purposes can be identified. The general purposes are:

i *Predicting or forecasting the future*
Based on the extrapolation of existing social economic and technological trends the near future is predicted assuming the trends will hold for the considered prediction period. In these models economic and growth patterns need to be considered endogenously.

ii *Exploring the future*
Different scenarios evaluate possible developments in the context of a "base case" or "business as usual" scenario. This approach is more based on assumptions than on extrapolating past trends and therefore can explore possible futures outside of current developments.

iii *Backcasting*

In backcasting, an advantageous future is envisioned by collecting expert opinion and consecutive consideration of necessary changes to achieve this vision.

The specific purposes according to [34] are firstly *Energy Demand Models*, that regard demand as a function of e.g. population, income and energy price changes. The second specific purpose are *Energy Supply Models* that focus on the technological side of energy systems and matching demand with supply and can incorporate economic aspects via optimization. Thirdly, *Impact Models* are distinguished that investigate the social, economic or environmental impacts caused by e.g. policy schemes or technological changes to the energy system. The last specific purpose mentioned is that of *Appraisal Models*. These allow comparison and evaluation of different options and their consequences.

Model Structure Regarding Internal and External Assumptions

Based on [37] a distinction between four different structural dimensions is made. The first of these is the degree of endogenization, i.e. how independent the model is from external assumptions. The second and third category regarding model structure are to which extent other (non-energy) sectors are included in the model and how detailed energy end-uses are implemented. Lastly, the detail with which energy supply technology is considered is important, as new technologies and changes in efficiency etc. need a certain amount of detail in order to be investigated adequately.

Top-Down and Bottom-Up

A summary of the comparison of top-down and bottom-up characteristics, depicted in table 2.1, is given by [34] based on the insights from [37, 38].

There is a trend towards closing the gap between top-down and bottom up models, newer versions of both incorporate more characteristics of the respective other, i.e. top-down models are increasingly detailed on a technological

Table 2.1: Characteristics of top-down and bottom-up models [34].

Top-Down Models	Bottom-Up Models
use an "economic approach"	use an "engineering approach"
give pessimistic estimates on "best" performance	give optimistic estimates on "best" performance
can not explicitly represent technologies	allow for detailed description of technologies
reflect available technologies adopted by the market	reflect technical potential
the "most efficient" technologies are given by the production frontier (which is set by market behavior)	efficient technologies can lie beyond the economic production frontier suggested by market behavior
use aggregated data for predicting purposes	use disaggregated data for exploring purposes
are based on observed market behavior	are independent of observed market behavior
disregard the technically most efficient technologies available, thus underestimate potential for efficiency improvements	disregard market thresholds (hidden costs and other constraints), thus overestimate the potential for efficiency improvements
determine energy demand through aggregate economic indices (GNP, price elasticities), but vary in addressing energy supply	represent supply technologies in detail using disaggregated data, but vary in addressing energy consumption
endogenize behavioral relationships	assess costs of technological options directly
assumes there are no discontinuities in historical trends	assumes interactions between energy sector and other sectors is negligible

level while many bottom-up models begin to incorporate endogen descriptions of the economic environment the bottom-up optimization is embedded in. The bottom-up approach can be distinguished in prescriptive and descriptive models. Prescriptive models make use of only of the most efficient technologies or achieve this by means of cost optimization. Descriptive models on the contrary aim at modeling the technology mix that society would implement under certain policies. Descriptive models are less optimistic than prescriptive models and have a higher degree of prediction as they try to represent possible paths resulting from actual decisions. [37]

Methodology

This account of model methodology is based on a summary of methodologies compiled by [34] with reference to [37–41].

  i *Econometrics Models*
    These models use statistical methods to make predictions regarding the future based on past market patterns using aggregated data. A common use is the investigation of interactions between energy system and economy.

  ii *Macro-Economic Models*
    This approach is centered on the sector interactions inside the economy. It is many times used to explore possible futures with assumed parameters instead of relying on realistic scenarios. As in econometrics models technologies are not captured in detail.

 iii *Economic Equilibrium Models*
    Economic equilibrium or resource allocation models are focused on long-term impacts. The energy sector is considered as part of a surrounding economy. A partial equilibrium model takes into account only a part of the economy, e.g. the energy sector considering equilibrium between demand and supply, whereas general equilibrium models consider an equilibrium state in all markets at the same time. This means general

equilibrium models allow for interactions between different markets. The path towards the equilibrium and therefore transition costs are not a model output.

iv *Optimization Models*

These models use mathematical optimization to achieve an energy system that supplies demand at e.g. lowest cost. As a tool it is used by companies and local authorities to take investment decisions as well as by governments for national energy planning.

v *Simulation Models*

A simulation model seeks to capture the relevant part of real world behavior of a system inside a model and allow for experiments on the model that enable drawing conclusions about the real world system. It is called static if it regards system operation in a single period and dynamic if the system evolves over time [42].

vi *Tool Boxes*

Also called spreadsheet models, refers to bottom up software packages for model generation, often containing example models that can be adapted to ones needs [37, 38].

vii *Backcasting Models*

See: *Purposes of energy models* on page 16, *iii. Backcasting*

viii *Multi-Criteria Methodologies*

Multi-criteria methodologies are applied in order to consider other goals besides economic efficiency, e.g. reduced greenhouse gas emissions and economic efficiency.

Mathematical Approach

Three mathematical means with which a result can be achieved in case of optimization, i. to iii. below, are explained by [34]. Two additional techniques, namely fuzzy logic and multi-criteria are mentioned but not explained further. Nonlinear and stochastic programming are named by [43].

i *Linear Programming (LP)*

"Linear programming is a practical technique for finding the arrangement of activities, which maximizes or minimizes a defined criterion, subject to the operative constraints (Slesser, 1982). All relationships are expressed in fully linearized terms. Linear Programming can be used, for instance, to find the most profitable set of outputs that can be produced with given type input and given output prices. The technique can deal only with situations where activities can be expressed in the form of linear equalities or inequalities, and where the criterion is also linear. That is, if x1 and x2 are inputs and y is the output, the technique is only applicable if their relationship is of the form y = ax1 + bx2. LP is a relatively simple technique which gives quick results and demands little mathematical knowledge of the user. Disadvantages are that all coefficients must be constant and that LP results in choosing the cheapest resource up to its limits before any other alternative is used at the same time for the same item (World Bank, 1991). Also, LP models can be very sensitive to input parameter variations. This technique is used for almost all optimization models, and applied in national energy planning as well as technology related long-term energy research." [34]

ii *Mixed Integer Programming (MIP)*

"Mixed Integer Programming (MIP) is actually an extension of Linear Programming which allows for greater detail in formulating technical properties and relations in modeling energy systems. Decisions such as Yes/No or (0/1) are admitted as well as nonconvex relations for discrete decision problems. MIP can be used when addressing questions such as whether or not to include a particular energy conversion plant in a system. By using MIP, variables that cannot reasonably assume any arbitrary (e.g., small) value –such as unit sizes of power plants– can be properly reflected in an otherwise linear model. (World Bank, 1991)." [34]

iii *Dynamic Programming*

"Dynamic programming is a method used to find an optimal growth path. The solution of the original problem is obtained by dividing the original problem into simple subproblems for which optimal solutions are calculated. Consequently, the original problem is than optimally solved using the optimal solutions of the subproblems." [34]

iv *Fuzzy Linear Programming*

This methodology allows using the same principles as linear programming while either the decision varialbes, the parameters or both are fuzzy, i.e. contain fuzzy numbers according to [44, 45]. This technique is used when input data is not known with certainty and thus allows for optimization with imprecise data [46]. A summary of different uses and methods to solve fuzzy programming problems is given in [47].

v *Multi-Criteria Optimization*

In multi-criteria or multi-objective optimization an optimal solution regarding more than one objective function is achieved. Usually there is not one optimal solution, but various solution which can not be improved regarding one of the criteria without worsening another. Multi-objective optimization is concerned with finding these solutions, as well as helping decision makers choose the preferred optimal solution according to individual preferences. [48, 49]

vi *Nonlinear Programming*

In contrast to linear programming the constraints and objective function of the problem description can be nonlinear.

vii *Stochastic Programming*

Stochastic programming aims at solving decision problems if data is uncertain. The techniques allow to take decisions optimizing the probable outcome by adding additional stages to the objective function of the optimization problem that allow optimizing the expected value for all realizations of the random event. [50]

Geographical Coverage

Many models can be used on different geographic levels, yet usually models have been created and are most suited for a specific set up. Van Beeck [34] divides these different geographic dimensions, in order of decreasing in size, in global, regional, national, local and project level. Regional thereby refers to international regions, e.g. Europe, and local to a sub-national level, e.g. a state level. The project level typically refers to district level geographical coverage, inside a specific project scope and context.

Sectoral Coverage

Any coverage of sectors is possible, from energy, or even electricity only, to multi-sectoral models that encompass the economy as a whole. The sectors that are not the focus of the model usually are taken into account in a very simplified manner. [34]

Time Horizon

Relevant time scales are identified by [38] to be less than five years for short term, from three to 15 years for medium term and more than 10 years for long term models. As one notices the time spans of this definition overlap, indicating that besides the number of years also the assumptions, e.g. consistency of past trends or the accounting for transitional effects, play an important role and help assigning one of the descriptions to the specific model. Grubb et. al link short term models to a rather transitional view and long-term to an equilibrium view [38].

Data Requirements

Data requirements span a broad range. From monetary values for technology costs and technology characteristics, such as efficiencies as well as detailed demand curves especially in bottom-up models, to more aggregated data regarding social development and the economic environment for top-down models. Generally detail decreases and aggregation increases with greater time horizons and geographical coverage. [34]

The above overview of the different model aspects gives an impression of the complexity and variety of possible model configurations. Exemplary for the vast number of models is [51], a review describing 125 tools for the scrutiny of electric vehicle integration into power grids and related applications.

2.2 Uncertainty in Energy Systems

Many of the input parameters necessary for the use of the before mentioned optimization models are not known. One consider a hypothetical situation of low complexity in which one wants to optimize a local energy system to be commissioned next year. It becomes clear that e.g. the total cost a certain technology such as a gas turbine or gas engine will cause during its lifetime is not known a priori. This is because it depends partially on unknown parameters such as uncertainties during construction, probabilistic events such as equipment failure and the future development of e.g. natural gas prices. In general uncertainty in a model becomes more difficult to evaluate, the more complex the model is, as the interaction between different uncertain parameters is more difficult to asses in more complex systems. In this section different origins of uncertainty in ESO are assessed and an overview on the literature concerning the prediction of technological development is given. The classification of uncertainty based on [52] and different concepts of probabilities as well as the human ability to interpret these are summarized. The section closes with a description of the most common modeling approaches to account for uncertainty in energy system models.

2.2.1 Origins of Uncertainty

Concerning the origins of uncertainty in ESO, De Carolis et al. [53] name parametric and structural uncertainties, based on [52, 54]. Parametric uncertainty is the uncertainty described in the example at the beginning of this section, it is the uncertainty surrounding the exact value of ESO input parameters. Understanding a mathematical model as a simplification of reality according to [55] based on the definition by Beck and Arnold [56] this type of uncertainty concerns the input parameters. The other dimension, structural uncertainty, is present in the description of interaction by mathematical means, that are used to produce the model output based on the input parameters. A famous example of structural uncertainty is the World2 model that lead to the "Limits to Growth" study in 1972 [57]. It is controversial if the assumptions on exponential and linear growth models are appropriate as well as if the accounting for market effects such as replacement of scarce and therefore expensive resources has been done sufficiently [54].

Examples of parametric and structural uncertainty are listed in table 2.2.

Table 2.2: Examples of parametric and structural uncertainty in ESO.

Parametric uncertainty	Structural uncertainty
Technology cost	Validity of linearity assumption
Efficiency of a process in the future	Neglect of part load behavior
Cumulative energy demand	Market behavior (e.g. rationality assumption)
Time distribution of a demand	Neglecting losses
Characteristic generation time series of renewable energy	Not including specific technologies in the portfolio

Another type of uncertainty, namely deep uncertainty, is the research subject of Lempert, Popper and Banks which they describe as:

"..where analysts do not know, or the parties to a decision cannot agree on, (1) the appropriate conceptual models that describe the relationships among

the key driving forces that will shape the long-term future, (2) the probability distributions used to represent uncertainty about key variables and parameters in the mathematical representations of these conceptual models, and/or (3) how to value the desirability of alternative outcomes." [54].

Deep uncertainty, that is the inability to describe the uncertainty to the satisfaction of all stakeholders, can therefore concern structural (1) and parametric uncertainty (2) as well as the derivation of decisions based on the inability to agree on the value of specific outcomes (3). In this work the main focus lies on the influence parametric uncertainty has on the model output and the uncertainties' influence on the ability to interpret results.

2.2.2 Prediction of Technological Innovation and Development

One important aspect of parametric uncertainty is the development of technologies in the future. For companies seeking to improve their competitive market position, knowledge about future technological developments are valuable. Models used by research institutions, companies and government to draw conclusions from possible future developments as mentioned in section 2.1, are in need of a variety of input parameters. Since most of these models aim at an optimization under economic aspects, one important parameter is the expected future development of technology cost and performance and possibly the emergence of new technologies. Therefore the prediction of technological innovation and development has been investigated widely since the nineteenth century. This has been done with the aim to predict scientific and technological innovation and their interactions with economy and society in order to take investment decisions, predict future costs or to overcome contemporary and future obstacles [58–61]. Jantsch [59] was the first to compare and differentiate various methodologies for technology forecasting. He summarizes the state of technology forecasting from twelve OECD members states in the years 1965 and 1966 while tracing back the beginnings of this discipline to the 1940s. One of the important merits of [59] is the definition of the terminology and the distinction between the rather uncertain forecast and the certainty implying term prediction. Yet, in most of the analyzed literature, no clear difference is made between the two terms.

Literature concerning the forecasting of technological development states that the topic of research is subject to uncertainty [59, 62–65]. In the following some of the more important methodologies which are currently used to predict technology developments are briefly explained to give an impression of the methodological aspects and limitations of the different approaches. A classification of techniques and methods for the analysis of future technologies is given by [66]. The most important methodologies for the prediction and exploration of technological innovation and development are

1. Expert Elicitation including Delphi Method, Foresight Exercises, interviews, cross-impact analysis

2. Trend analysis by experience respectively growth curves, precursor analysis and long wave analysis

3. Bottom up cost analysis

2.2.2.1 Expert Elicitation

Expert opinion methodologies are based on harvesting the knowledge about a certain topic from experts in the field. The Delphi method collects information based on discussions between groups of individuals with the goal of a "consensus based response" [54]. It has been first applied and developed in the 1950s by RAND researchers, with the reports regarding the experiments around an estimation of necessary bombing of U. S. industrial targets with atomic bombs, published in 1962 [67]. The procedure is based on asking experts on their estimate of a specific question and make material requested or used by one expert available to the other experts in consecutive rounds. During the process care

is taken not to bias an experts opinion. For this reason direct confrontation between the experts is avoided. Only the indirect interaction of supplying information to all, that at least one of the others or the experimentalist deems to be necessary in order to make an informed guess about the question at hand, is allowed. The report by Dalkey and Helmer specifically names the desire to use the Delphi process as a possible substitute for empirical evidence if none can be obtained [67]. This is the application the Delphi method has been used for in the case of seeking to predict technological change, as e.g. [65, 66, 68]. For predictions about the far future Lempert et al. criticize the consensus aspect of the methodology, as many different outcomes on technology development are equally possible and a consensus necessarily neglects many plausible alternatives [54]. [65] reports inconsistencies between the given expert opinions on component and system level developments.
There are similar elicitation methodologies that differ from the Delphi method. An example for expert elicitation based on a combination of supplying background information, discussion of this information and elicitation of desired values in the form of interviews for the case of water electrolysis is given by [62]. The process differs from the Delphi procedure outlined above, as no information that has been requested by one expert is given to the others and consensus is not an aim of the procedure. Instead the use of the arithmetic mean of the individual predictions is proposed [62].

Foresight exercises are a group exchange of ideas in the form of "normative and positive views of the future" with discussions and the exchange of ideas from political figures and experts itself being the focus of the process, therefore allowing for direct interaction between participating parties [54]. A generic foresight framework for a 10 to 20 year foresight time horizon is given by Voros, who emphasises the need to inform the participants of the methodology and its goals beforehand, as otherwise the exercise is not taken seriously and compared to "crystal balls" [69]. It is also identified as an element of strategic thinking, as it has a creative, explorative nature of

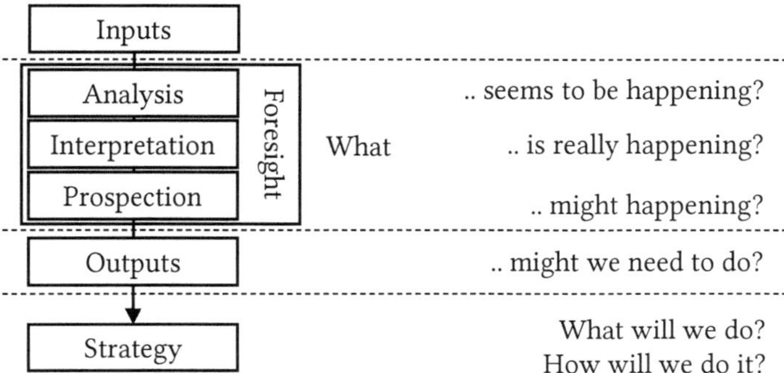

Figure 2.2: The foresight framework in question form based on [69]

uncovering options for future development [69]. A graphic representation of the foresight process based on [69] is depicted in fig. 2.2. One example of the application of foresight on a governmental level is the application by the German Federal Ministry of Education and Research (Bundesministerium für Bildung und Forschung) that has since 1992 engaged in Delphi studies which were later replaced by foresight exercises in two cycles, from 2007 to 2009 and 2012 to 2014. These were concerned with the analysis of technological trends and the identification of seven future fields and consecutive discussion with representatives from society and politics. In the second cycle, the basic assumption was that technological and social innovation is derived from the combination of technological possibilities and social necessities. The current foresight process evaluates trends and topics on a continuous basis in order to identify and discuss future relevant topics. [70]

Cross-impact analysis is based on experts evaluation of the likelihood of events both independently of and dependently on one another. The experts give relative likelihood scores to different statements, that are prepared before-hand by the experimentalist. In a second step the experts are asked to give a conditional probability of one event coming true depending on another event.

A comparison of cross-impact analysis with the Delphi method is performed by [68]. Scapolo and Miles point out the difficulty of implementing a cross-impact analysis for complex topics and report the impression of inconsistency between given answers, perceived by the participating experts. Also a general lack of practical advice on how to put an expert elicitation into practice is noted despite wide-spread application of some of the methods. [68]

2.2.2.2 Trend Analysis

Trend analysis according to [66] includes precursor analysis, long wave analysis and trend extrapolation in the form of growth curve fitting and projection as well as trend impact analysis. A precursor by definition of the Cambridge dictionary is "*something that happened or existed before another thing, especially if it either developed into it or had an influence on it*" [71]. Precursor analysis makes use of observed incidents, that offer information about possible future events. One important application of precursor analysis is safety analysis e.g. for nuclear power [72] or spaceflight [73]. The safety design of nuclear power plants is based on redundant systems and probabilistic safety analysis which, due to the combined probabilities of multiple system and component failures, leads to very low expected core damage accident rates [72]. Precursor analysis as performed by Ballard [72] can show that such events must be expected much more frequently than assumed during system design, which is also backed by empirical evidence showing that a core-melt accident can be expected with a probability of almost 70% within each ten years on a world-wide scale, and with a 50% probability within the next 25 years in the U.S. [74]. In the research regarding technological change precursors are used in order to identify lead-lag relationships between different technologies, that allow to conclude on the development of one technology based on that of the other, leading technology [75]. Long wave analysis seeks to identify long term cycles in past development [76]. Long wave theory assumes forty to sixty year cycles of technological development and surrounding social evolution [77]. An overview over different schools of thought on the interpretation of long waves and the reason for cyclic technological development can be found in [77]. Another form of trend extrapolation is growth curve fitting and projection, which is based on the analogy of growth between organisms and technology development and adaption [78]. It has been shown for the development of a range of technologies to fit the logistic growth function, leading to first noticeable successes in the prediction of technological change [75]. A comparison of different growth curves for energy technologies, namely oil production, oil field discovery and nuclear power production in the U.S. is presented by [79]. It reveals uncertainty even if an underlying growth curve assumption is correct, as different growth curves can have similar fit qualities for past data, but significantly differ in their predictions for the future. Growth in technology implementation is closely linked with its cost as the application of experience or learning curves reveals. Learning curves were originally observed in individuals who were learning a task while performing it, showing that the task was performed more efficiently, i.e. faster with learning. Also it was shown that the velocity with which a task is performed more efficiently decreases with increased learning. Later these findings were found to be applicable also for groups and organizations as a whole. [80]

The experience curve is a generalization of the learning curve, which gave the initial hypothesis to apply cost reductions to cumulative industry production instead of relating only the time to perform a certain task to the time spend learning within an organization [26]. The basic assumption that is supported by the empirical evidence is that the cost of a certain product decreases by a certain percentage for each doubling of cumulative produced volume of that product in the whole industry [26]. Experience curves as well as their contemplation in combination with growth curves are a broad field of investigation. A recent review of experience curve application on electricity generation technologies is performed by [81]. The author finds that uncertainty regarding specific learning rates, i.e. the rate with which costs decrease for each doubling of capacity, in the future exist as there have been fluctuations in the past and urges to take into account other factors that might influence the future development of learning rates. Taking into account unexpected future events in the extrapolation from past data is what Trend Impact Analysis is aiming to do [82]. The application of Trend Impact Analysis is based on firstly evaluating a trend from the past and extrapolating it into the future, given no unexpected events occur. Secondly it includes asking for expert judgment to identify future events that, if they occur, cause a different unfolding of future development than predicted by the trend. For these events the probability of occurrence is evaluated over time, resulting in a range of possible future developments. [82]

2.2.2.3 Bottom-Up Cost Analysis

Bottom up cost analysis is based on evaluating the costs of the individual components and production processes in order to calculate the total cost of a system and possible cost decreases in the future. The analysis usually takes into account raw materials and then subsequently prices in the cost of manufacturing and assembly as well as risks, taxes and installation costs. Bottom up cost analysis can include the use of other cost estimation techniques for subsystems of a technology, such as the use of experience curves to estimate subsystem cost reduction potential. Examples of bottom up cost analysis with the prospect to estimate possible future cost reductions are e.g. [63, 64, 83]. Bottom up cost analysis can uncover uncertainties in the future that conventional use of experience curves cannot and allow for an identification of the possible sources of future cost reductions [84].

2.2.3 Uncertainty and Probability in Energy System Models

Morgan and Henrion [52] summarize the different types of uncertain quantities regarding a model and how uncertainty in each of these can be accounted for in a policy model. Their work, being focused on policy analysis, is applicable also to ESO, as these models often serve a more or less explicit use as a basis for decision making in policy. Table 2.3 shows the different uncertain quantities and how uncertainty should be treated according to Morgan and Henrion.

Table 2.3: Uncertain quantities in ESO models based on [52]. Switchover refers to a discreet, non-continuous variable, compared to continuous parametric variables.

Type of quantity	Example from ESO	Treatment of uncertainty
Empirical parameter or chance variable	Process efficiency, commodity cost	Probabilistic, parametric or switchover
Defined constant	Joules per kWh	Certain by definition
Decision variable	Plant size, emission cap	Parametric or switchover
Value parameter	Discount rate	Parametric or switchover
Index variable	Time period	Certain by definition
Model domain parameter	Geographic region, time horizon, time increment	Parametric or switchover
Outcome criterion	Total system cost, optimized plant size	Determined by treatment of its inputs

Empirical quantities are properties of the modeled system that can be measured, while decision variables are those variables the decision maker can actively influence [52]. This adds decision parameters apart from the plant size which is a typical decision parameter from the optimizers point of view. Also boundaries as e.g. upper limits for technologies such as wind power plants are decision variables, if the upper limit is not defined by physical means, but rather by decisions made in politics and society. Regarding table 2.3 it becomes clear that for empirical parameters and chance variables a probabilistic treatment is recommended, while the influence of all other uncertain parameters on the model results should rather be analyzed by means of parametric sensitivity analysis. As the performance of a parametric sensitivity analysis is rather straight forward, in the following the probabilistic concepts and the basics of probability distributions are explained briefly. [52] gives an overview over the most important aspects of the probabilistic treatment of uncertainties, one of the important distinctions is between the frequentist and the Bayesian view of probability. The classical or frequentist approach is to perceive the probability of a specific outcome as the value to which the experimental count of observations of that outcome in relation to the number of trials converges with increasing number of trials. For most parameters regarding energy system modeling, it is impossible to actually perform experiments and therefore retrieve a frequency based estimation of an event's probability. This is due to the fact, that the real world system being modeled is usually one of a kind and events to be evaluated are unique such as innovation, which, once achieved within a certain time-span, can hardly be undone in order to perform the next experiment and see how long it takes this time. This is the reason the Bayesian, personalistic view on probability is more useful for the purpose of describing the probabilities of empirical variables in policy models. The personalistic view assumes that the probability one has to assign to an event does not only depend on the event itself, but on the information available. This means that probability is a rather personal thing, that depends on who defines it as much as on the event itself. This might seem arbitrary, yet with more information the probabilistic assessment of an individual observing a coin toss will converge with the frequentist interpretation. Also it is important to note that also personalistic probabilities must object to the same axioms as objective or frequentist probabilities. This means that probabilities must be mutually exclusive and the sum of the probability of all possible events must be one. [52]

Humans use heuristics to judge uncertainty and probability that are usually satisfactory in everyday life, while at the same time are challenging for scientific practice [52, 85]. [52] summarize some of the errors in human judgment, largely based on studies performed by Kahneman and Tversky, and relate these to assigning probabilities. These fallacies are listed below.

- Availability [86, 87]
  Something is perceived as more probable because it is more accessible to the mind of the judge, either due to recency or because of prominence.

- Representativeness [88–90]
  There are two important aspects to this. Either the behavior of the empirical probability for big sample sizes is expected also in small samples or a more descriptive, i.e. more easily imaginable scenario is believed to be more probable, even if this probability is against a known base rate.

- Anchoring and adjustment [52, 90]
  A starting point for an approximation is given, from which e.g. the extremes of a probability range are adjusted. This adjustment is usually too small, resulting in a too narrow probability range.

All of these will have influence on someone trying to give a subjective probability for future events. Especially the aspect of representativeness in the context of ESO and scenarios is striking, as a combination of probabilities in a graphic scenario are judged likely, even though the combined occurrence of events, e.g. a certain cost for all the considered technologies, has a probability that is some orders of magnitude smaller. To the experience of the author, if someone is asked how likely it is that photovoltaic cost will decrease by 50% until a certain year and than an ESO scenario, e.g. called "solar cost decrease", is presented, the other person will in most cases judge both cases with a similar probability, even though all the other assumptions, necessarily present in the scenario, lower the probability of the combined event significantly.

The research outlined above has mainly been performed on university students and "normal people" i.e. non experts, hence one important question is if experts perform better [52]. Morgan and Henrion give an overview over the relevant literature on the subject before 1986 without reaching a clear conclusion, also due to the fact that for unique events there were few empirical studies that allow for a control of the predictions quality. In his 2005 work "Expert Political Judgment" Tetlock gives an overview over the results of his long term research until 2004 [6]. The study was designed to collect long term predictions about real, verifiable, global events and compare experts precision with that of ordinary people, algorithms and chance. The results show that there are only small differences in the performance of experts and non-experts. More important than the distinction between not-experts and experts is the way of thinking. People who are thinking of many different aspects and are open for contradictory information while making a prediction, perform better than those who think about the big concept of the forecast and have difficulty incorporating thoughts that challenge their beliefs [6]. In a follow-up book from 2015, again based on numerous publications in scientific journals, Tetlock et al. uncover the underlying habits and traits, that distinguish people who continuously make better forecasts than others [91]. The methodology of how to identify and improve the performance of consistently successful forecasters are described in [92]. The findings show that exceptional forecasting results are not to be expected from any expert, but that there is instead the need to identify, train and combine the right people in order to consistently achieve more reliant forecasts. This research from Tetlock et al. suggesting that prediction under certain circumstances can be performed quite reliably is only valid for short and medium term forecasts up to 1.5 years in advance [6]. Tetlock states that forecasting results for time frames above 5 years converge to chance predictions, even for someone capable of performing well on short term forecasts [6].

2.2.4 Taking Uncertainty into Account

The above discussed presence uncertainties, especially regarding future parameters of energy systems, have lead to a variety of approaches in order to gain insights despite not knowing many of the important parameters for sure. In this section some of the more important approaches in the field of ESO are briefly discussed and a literature based evaluation of each is given. The methodologies discussed herein include scenario approach, sensitivity analysis, robust optimization and stochastic optimization. Based on table 2.3 some of these techniques are only applicable for the investigation of certain kinds of uncertain variables.

2.2.4.1 Stochastic Optimization

According to the definition from WolframMathWorld^{TM} [93], stochastic optimization is a *"[...] minimization (or maximization) of a function in the presence of randomness in the optimization process. The randomness may be present as either noise in measurements or Monte Carlo randomness in the search procedure, or both"*. For ESO this randomness as outlined above is present in empirical and chance parameters that can be described with probability distributions as well as discreet probabilities. For the purpose of investigating ESO models we differentiate between decision focused and exploration focused stochastic optimization.

The decision focused approach centers around the idea of identifying the probabilities of future developments, incorporate these in the model formulation and solving the problem such that the optimal decision based on the state of knowledge concerning the uncertainties is achieved. An introduction to the topic as well as an overview over different applications and specific challenges is given by [50]. The result of a decision focused optimization in the case of energy system are, e.g. specific capacities of different technologies that need to be installed and how much the resulting system will probably cost. The advantages of this approach are mainly clarity for the decision maker and fast computation. Literature containing examples of applications is mentioned by DeCarolis et al. [53].

If a decision maker seeks to understand the possible ranges of optimal solutions, it can be helpful to actually compute the probability distributions of optimal system layouts depending on the probability distributions of the input parameters. This is achieved with the exploration focused approach. The exploration focused approach does not yield single figure results, but instead probability distributions for the decision variables. According to the definition of stochastic optimization by Morgan et al. it can be described as *"[...] running the analysis over and over again on a fast computer, using different input values, from which it is possible to compile the results into probability distributions"* [94]. The advantage is, that the whole range of possibly optimal solution and how they are weighted respective to one another is available for deliberation. Further analysis can be performed and planning decisions can be taken with more information. The greatest disadvantage is the computational effort. Even though current computers make it possible to run ESOs in a reasonable amount of time it can be demanding and time consuming to perform several hundred optimizations [95]. To the knowledge of the author there is no relevant literature concerning the use of this exploration focused approach in energy systems optimizations except [95, 96], the findings of which are the basis of this work.

2.2.4.2 Scenario Approach

Scenarios have been described as *"...plausible and often simplified descriptions of how the future may develop based on a coherent and internally consistent set of assumptions about key driving forces and relationships"* [97]. Scenario planning allows to synthesize different plausible futures and therefore to incorporate the diverseness and uncertainty of possible futures [54]. Lempert et al. point out that the scenario approach is especially useful to promote an understanding that the future, especially the long term future, may be very different from today [54]. This aspect is also stressed by Rasmussen [98], who focuses on how to tell a story that supports the understanding of a scenario and improves the impact of the related research. An overview over the use of scenarios in the historic context, beginning in military strategic planning, is given by Rounsevell and Metzger [99] together with an overview over different methods of scenario storyline development. These methods have been developed, because an important aspect of scenario investigations is the consistency of future developments within the scenario storyline. This has also made clear by the definition in the Millennium Report [97].

In order to assure a consistency between the different assumptions various scenario generation techniques have been developed. DeCarolis et al. summarize the different aspects to scenario generation and reference relevant literature[53]. Lempert et al. state that the different scenarios generated should be plausible but diverse. This includes, that an error on the site of considering also versions of the future (for simplicity: futures) that might be implausible is preferable to the loss in robustness of the analysis if extreme or unlikely scenarios are not included, as this will lead to decision makers being surprised by the unfolding of reality [54]. Hughes [100] gives an overview over the complex development and the different branches of scenario development and the concerned individuals and research institutions in form of a scenario family tree. He also points out that even tough dystopian and utopian scenarios may have some informative value of different extreme futures, more value can be derived from a well designed scenario if uncovering causes and effects gives the decision maker the possibility to improve her reasoning by considering additional possible results from different courses of action or inaction [100].

2.2.4.3 Sensitivity Analysis

The practice of sensitivity analysis usually addresses small variations, coherent with the notion that a modeler does not possess knowledge in advance how even small variations of the input parameters will influence model results. A factor, i.e. an input parameter, or multiple factors, are varied and the change in model result parameters is recorded. One approach that is used many times in the analysis of models and technical systems is a so called "one factor at a time" variation, even though this reduces the investigated parameter space and neglects factor interactions [101]. Especially for large models global sensitivity analysis approaches are more effective [102]. This fact is well known also in the case of physical experiments where the alternative global approach is known as *Design of Experiment.* Yet many times also for experiments one factor at a time variations are used, even though there is a real cost associated with each experiment [103]. Siebertz et al. argue that it is very expensive to use one factor at a time analysis, as the amount of information gained in each experiment is not maximized [103]. An overview over various global sensitivity analysis approaches is given by [104]. The suggestion by Siebertz et al. to use probability distributed Monte Carlo analysis as a possible sensitivity analysis approach effectively closes the gap to the above described exploration focused stochastic optimization.

2.2.4.4 Robust Optimization

Robust optimization is useful when the input parameter probability distributions to a real life optimization problem with uncertain data are not known [105]. It has been defined as the search for an optimum under the condition of optimizing the original objective, e.g. total cost, while at the same time ensuring only small variations of the objective when subject to input parameter variations [106]. The mathematical basis for robust optimization have been published in the 1970s [107] and has been applied in many different fields [105]. Applications in energy systems include unit commitment investigation [108] and long-term planning for domestic energy systems [109]. Advantages of robust optimization include that no input parameter distributions are needed [105, 109]. The need for robustness can lead to very conservative solutions as even the most improbable worst case combinations are considered. For this reason the use of uncertainty sets is common, which allow significantly reducing this effect while ensuring that the robustness constraint is "almost never" [105] violated. This approach reduces the advantage of using robust optimization as it necessitates an approximation of the probability distributions of the uncertain problem aspects.

Chapter 3

Research Objectives and Methods

3.1 Research Objectives

ESO is a valuable tool for industry, politics and society to derive information for investment and policy decisions. With optimization results an educated guess of a possibly close to optimal amount of electricity and heat generation technologies can be determined and policies to achieve a sufficient implementation and installation can be derived. In order to design an optimization model, assumptions for future costs and efficiencies of the considered technologies must be made. While the use of ESOs for these purposes is an improvement compared to other decision making basis such as e.g. yearly or monthly energy balances, there are still shortcomings regarding the planning value. As apparent from the literature presented in section 2.2 on page 24, uncertainty is undoubtedly present when trying to predict technology parameters for the future. Yet this uncertainty is currently mostly left unconsidered and its effects on ESOs have not been characterized systematically. In order to allow for the possibility that ESOs will continue to improve policy and investment decisions these effects are investigated. Therefore the focus of this work is on the quantification of the effects uncertainty has on an optimal system layout compared with a scenario based on the means of stochastic optimization. An additional focus lies on determining the influence of the number of discretization points of stochastic optimization to determine information loss if small distributions are used, e.g. due to high model complexity and resulting computation times.

Effects on the optimization objective are considered, as well as the effects on the optimized parameters such as the optimal installed capacity of storage or electricity generation technologies. These effects are investigated with regard to the impact of the width of input parameter distributions representing the amount of uncertainty present and to the number of uncertain parameters in the system. It is investigated if systematic errors occur when uncertainty is present and neglected in the optimization. The presence of systematic errors makes a quantification on their effects for future research considering ESO necessary and allows to assess possible shortcomings of past research that does not consider the effects of uncertainty. Additionally a surrogate or meta model is derived and its results compared to those of the stochastic optimization according to section 2.2.4.1. This aims at lowering the computational expense for stochastic optimization and allow for more wide spread adaption of the methodology as the additional information achieved from stochastic optimization under uncertainty is retained.

3.2 Methods

In order to achieve the research objectives, different aspects are investigated according to fig. 3.1. In chapter 4 theoretical considerations regarding uncertainty in optimization problems are applied to uncover systematic errors. An exemplary ESO with low complexity is used to experimentally investigate the findings from the theoretical considerations and to evaluate the influence of the amount of uncertainty. The amount of uncertainty is varied regarding the number of uncertain parameters and the width of probability distributions of the input parameter sets. Considering these findings, the impacts on the interpretation of ESO results are evaluated. To further explore the impact of uncertainty a model representing the German energy system regarding heat and electricity consumption and their production as well as the inner-German transport sector is formulated. The details of the model are presented in chapter 5. The results are reported in chapter 6 and evaluated in chapter 7. The investigation is performed for three different emission reduction boundary conditions, namely a minimum emission reduction of 80 %, 95 % and 100 % compared to 1990 emissions from the energy sector without diffuse fuel emissions according to [3]. The model complexity is comparable to models that have been the basis of reports and scientific publications in the past. For the evaluation first stochastic and scenario results are reported and than compared. As interpretation and communication of stochastic optimization results are less straight forward than in the case of scenarios, in section 6.3 clustering and factor effect methodology are applied to gain additional information regarding the underlying structures of the optimized systems. To promote the use of stochastic optimization in future research, possibilities to use less optimizations are investigated. On the one hand the use of smaller discretizations of the uncertain parameters probability distributions is investigated regarding the effect of less optimization runs on the box-plots, especially result means and inter-quartile-range, as well as on clustering. Additionally the use of meta models based on factor effects to reduce computational demand is evaluated by comparison of the resulting distributions with the exact results. The investigation is followed by a discussion of the results in chapter 7. Therein advantages and drawbacks of stochastic optimization are considered in comparison to the scenario approach. The trade-offs when using less optimizations, as well as the methodological feasibility and added value of clustering and meta model use, are considered. The work concludes with a summary of the findings and the identification of future research demand in chapter 8.

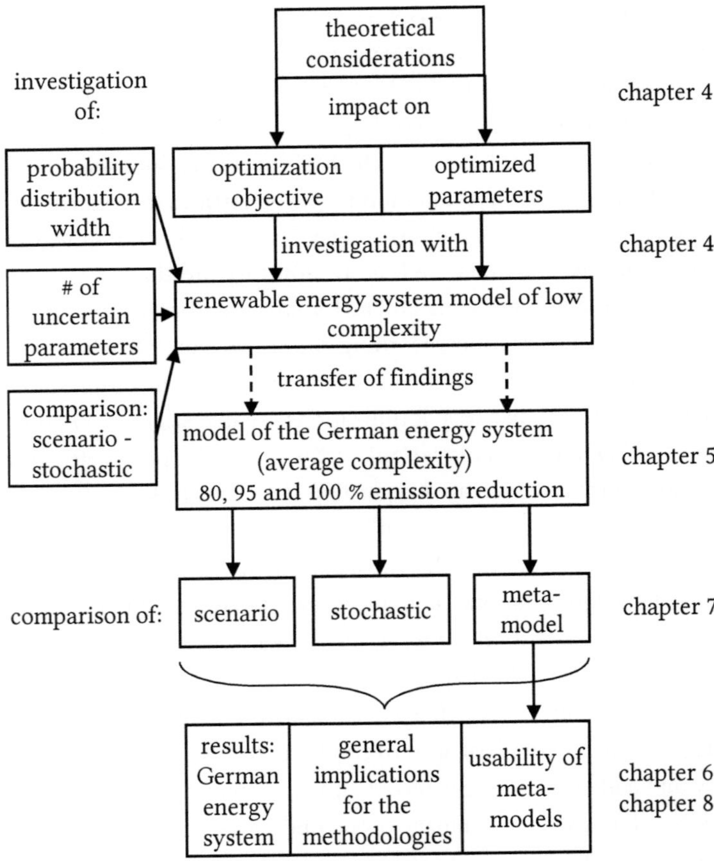

Figure 3.1: Investigation of uncertainties and corresponding chapters.

Chapter 4

Systematic Errors if Uncertainty is Ignored

It has been shown that systematic errors are to be expected if uncertainty is not taken into account in optimization under uncertain input parameters in general [110] and for the case of energy systems specifically [111, 112]. Madansky has shown that generally equal or better, i.e. more optimal solutions, result by explicitly solving an optimization for a random vector in the optimization problem formulation, compared to optimization considering the mean of the random vector [110]. Hobbs applies this to simplifying the description of available capacity of any technology in an energy system by an availability factor compared to taking into account the stochastic occurrence of outages [111]. Jensen's inequality states that, between the two points of intersection, the values of a line that intersects with a convex function are higher than the values of the convex function. Hobbs shows that, resulting from Jensen's inequality, the use of the above mentioned simplification using availability factors leads to an underestimation of operating costs as production from the most expensive units as well as necessary peaking capacity are underestimated [111]. In this chapter first the considerations that lead to the conclusions that systematic overestimation of the optimization objective occurs when uncertainty is not considered are laid out. Then a simplified ESO model is explained briefly. This model is used to evaluate the influence of the width of input parameter distributions and the number of uncertain parameters experimentally. After the model presentation key findings together with the most important conclusions are summarized. This chapter is based on the comparison of a cost optimization of a so called 'scenario approach' with a stochastic cost optimization. The 'scenario approach' herein represents a single optimization, that assumes the means of the underlying probability distributions as the value of the uncertain input parameters, following the definition from [95]. The stochastic optimization refers to the exploratory approach, a multitude of optimizations being used to discretize the underlying uncertain input parameter distributions.

4.1 Theoretical Considerations

The nature of the systematic error caused when optimizing uncertain energy systems without a proper representation or consideration of uncertainty is summarized by the following description of what happens when uncertainty is accounted for by including it in discretized form with stochastic optimization:

If an energy system is optimized considering uncertainty explicitly, the optimizer always mitigates the effect of cost increase and enhances the impact of cheaper technologies.

This effect leads to a scenario optimization that takes into account only the mean of an uncertain variable, generally overestimating the mean of the optimized system cost that results from the stochastic optimization. We can give a graphic illustration of that problem by a game of thought, where a solely photovoltaic (PV) and battery based energy system will, up to certain energy balance boundary conditions, react to different combinations of cheaper and more expensive batteries and photovoltaic panels. This simplified case is visualized regarding the effect of a variation in PV system cost in fig. 4.1.

A mathematical explanation is given when considering that for many of the common optimization problems the function that describes the optimization results over varying input parameters is a convex function [113]. This function

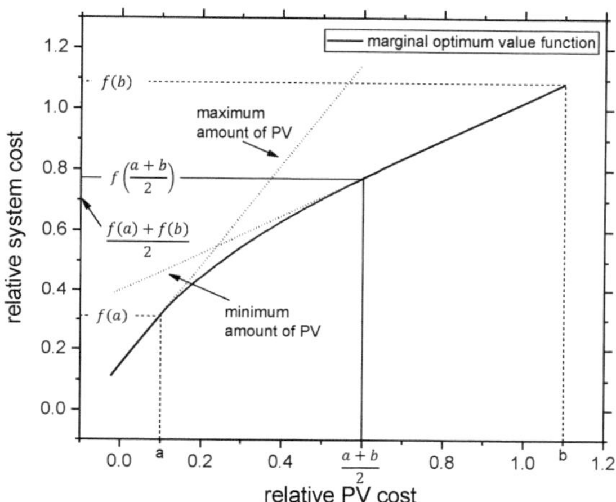

Figure 4.1: Schematic representation of the marginal optimal value function and the reversed form of Jensen's inequality for convex functions for a two-technology energy system optimized with regard to the total system cost.

is called the marginal optimal value function [114]. The optimization problems for which this is the case include the herein used linear optimization as well as quadratic optimization [115, 116]. This means that Jensen's inequality for convex functions [117], is applicable. For the case of minimization as schematically presented in fig. 4.1 the marginal optimal value function is concave, this results in the conclusion that generally an overestimation of the optimization objective will take place if the uncertainty is not considered in form of a distribution. For a simple distribution consisting of only two points, a and b this yields Jensen's reversed inequality

$$f(\frac{a+b}{2}) \geq \frac{f(a)+f(b)}{2}$$

for the total system cost. The expression describing this for the expected value E, a real-valued integrable random variable X and for a concave function g based on [118] is:

$$g(EX) \geq Eg(X)$$

Therefore the mean, i.e. the expected value of the optimization objective resulting from the optimization with distributed input parameters, is smaller than that of solving the optimization for the expected value of the input parameter distributions. Accordingly cost optimized systems overestimate the probable optimal system cost if such an approach is chosen.

4.2 Model to Investigate the Effects

In order to investigate the effect of Jensen's reversed inequality as well as how great the effect of different distribution widths and different numbers of uncertain parameters on the system is, a strongly simplified linear all renewable ESO model with the objective to minimize total system cost has been designed. In the model there are three renewable electricity generation technologies, photovoltaic, onshore and offshore wind power plants with corresponding generation time series. For electricity storage electrolysis, hydrogen storage, combined cycle hydrogen gas turbines, reversible solid oxide fuel cells and batteries are taken into account. The optimization yields the installed capacities of all the technologies that allow a satisfaction of electricity demand at the lowest cost. Each technologies cost is varied in a range of 10 to 110% of current cost, with the mean of the distribution fixed at 60% of current cost. As uncertainty can also apply to technology efficiencies, these are varied in reasonable ranges, with the means at the center of the extreme points of the distributions. Probability distributions for variable input parameters are created as a latin hyper cube (lhc) and transformed to represent normal distributions that are compressed in order to be contained by the boundaries of the interval. A detailed explanation of latin hyper cubes and the compression into a truncated normal distribution can be found in the appendix appendix A.1. and the numerical values and model details are included in appendix A.

4.3 Quantification of Effects

Different cases are investigated in order to quantify the effect width and number of distributions have. First, two optimizations with different width of the probability distribution, two and five percent, corresponding to 3.33 and 8.33% of the cost range, are performed and the results compared regarding the influence on the optimization objective and onshore wind capacity as an example technology. Then the number of parameters subject to uncertainty is varied and the resulting interquartile ranges, i.e. the range above which and below which each 25% of results lie, are analyzed for varying input parameter standard deviations (SD).

4.3.1 Effect of Distribution Width and Number of Uncertain Parameters

In fig. 4.2 the influence of distribution width on the output parameter distributions for the two above mentioned distribution widths are compared for optimally installed onshore wind capacity and the cost of energy, i.e. electricity in this case.

The histogram that is based on 1000 optimizations shows that, for the investigated cases, a wider input distribution leads to a wider output distribution. Especially the case of onshore wind and the peak at zero installed capacity reveals that output parameter distributions can be very heterogeneous. The results for the cost of energy which is directly calculated from the total system cost, dividing it by the total amount of energy demand, i.e. the sum of the demand time series, reveals a smaller mean cost at a greater input parameter

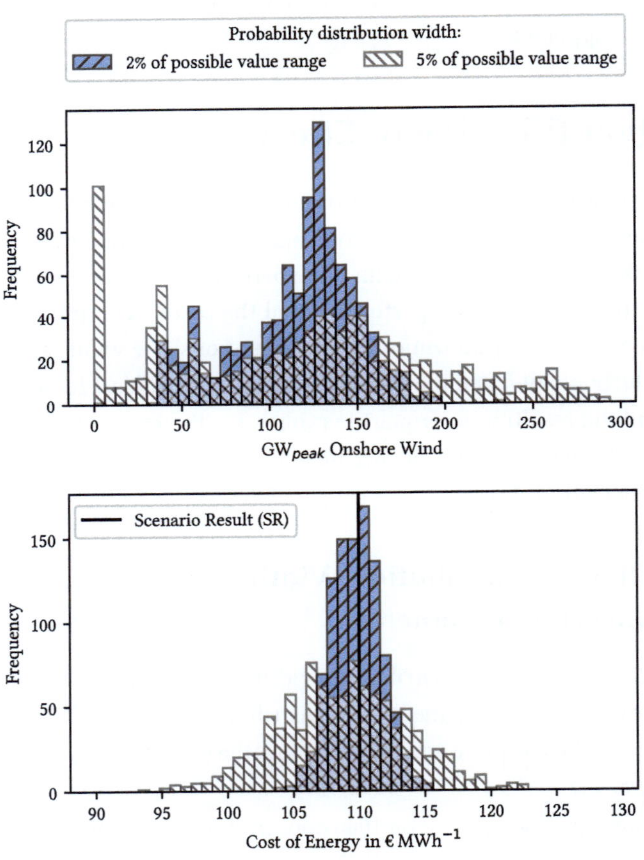

Figure 4.2: Influence of the distribution width on optimally installed onshore wind capacity and cost of energy.

standard distribution, namely 108.7 € MWh^{-1} in the five percent case and 109.8 € MWh^{-1} in the 2 percent case. That the scenario result, marked with a vertical line at 110 € MWh^{-1}, overestimates the mean of the distribution, is congruent with Jensen's inequality.

The effect the number of uncertain parameters has on the output parameters and the optimization objective is analyzed together with the influence of the distribution width, as also an interaction between both is possible. For the optimization objective, the distribution's mean, which is the expected value, is divided by the scenario result and plotted over the relative standard deviation, i.e. the standard deviation divided by the parameter range as defined above. The resulting graph is given in fig. 4.3. The black lines with descending transparency represent the increasingly present uncertainty, the solid line represents the result for the case where all parameters are uncertain. With greater standard deviations, greater deviations of the cost of energy in the scenario result. Also the number of parameters increases this effect, while the contribution differs greatly between the technologies, as visualized with the comparison of the cases in which only the battery capacity and only the onshore wind capacity are uncertain, with the case in which both are uncertain.

In order to be able to compare the results for all technologies on a similar scale, the relative interquartile range (IQR) is calculated by dividing the technologies installed capacity's interquartile range by the scenario result. This is again plotted against the relative input parameter standard deviation. Since the relative IQR of the results differs for the technologies these have been divided in two groups, represented on different scales in fig. 4.4. The general trend, that higher input parameter standard deviations yield broader output distributions, is discernible. The comparison of the lines with different transparency that mark the different input parameter uncertainties shows that also here a higher number of uncertain parameters results in a higher uncertainty regarding the output parameters. Even very small input parameter distributions, only shown for the case with uncertainty for all input

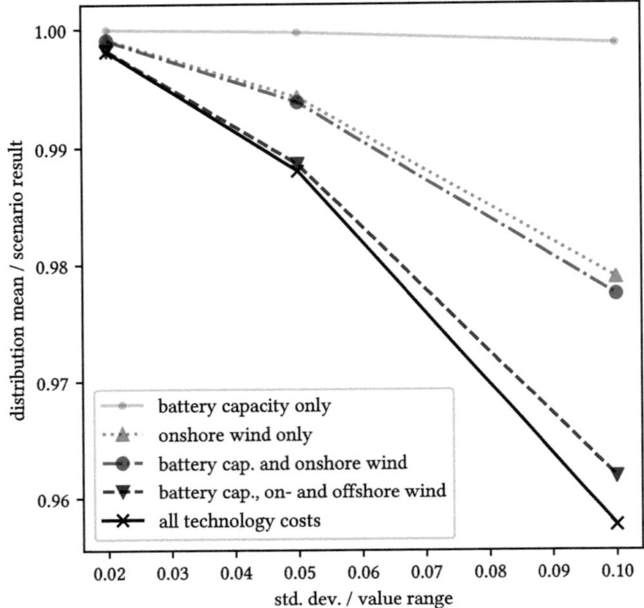

Figure 4.3: Influence of distribution width and number of uncertain parameters on the mean of the cost of energy normalized with the scenario result.

parameters in fig. 4.4, can lead to comparably big IQRs for technologies such as the optimally installed battery capacity in this example. In order to get a better understanding of the above mentioned general implications, the cases of onshore wind, in the upper plot, and PV, in the lower plot in fig. 4.4, are examined in more detail. For onshore wind, represented by the light green lines with small, round circles as symbols, the first trend can be observed by taking a general look at the slope of the corresponding graphs. The slope is positive. This means that higher standard deviations in the input parameters, depicted on the horizontal axis in normalized form, lead to a wider range of installed onshore wind capacities in the results, depicted by the normalized IQR on the vertical axis. Higher input variations thus lead to wider output IQRs. When comparing the light green lines with different transparency and line type, a conclusion about the influence of the amount of uncertain parameters on the installed capacity of onshore wind can be made. With increasing transparency less technologies in the investigated system are considered with uncertainty. The most transparent line, a solid line, is nearly parallel to the horizontal axis. Therefore the presence of uncertainty only with regard to the cost of battery capacity has little influence on the width of the output distribution of optimally installed onshore wind capacity. This is almost independent of the width of the input distribution, as even for 10 % distribution width there is no significant increase in the results IQR. Considering only uncertainty of the technology itself, in this case onshore wind, changes this. For this case a 10 % input distribution yields an IQR greater than 100 % of the scenario result. Adding also battery capacity cost as uncertain does not change this, the lines overlap. When additionally considering uncertain offshore wind cost, again a jump in IQR can be discerned. In this case 10 % input standard deviation yield an IQR of almost 175 % the scenario result. Adding all other technology costs uncertainty does not have a big impact, again the resulting lines overlap.

For PV, depicted in the lower plot with yellow triangles, similar trends can be observed. The most notable difference is, that in this case each addition of uncertainty for a technology yields an increase in the IQR of installed PV. There are no cases where there is little to no change even though uncertainty is added. Interestingly in the case of PV the presence of uncertainty in battery capacity and onshore wind cost at the same time has less impact than the presence of uncertainty only regarding the onshore wind cost. This becomes clear by the comparison of the dotted and dash-dot line. The general trends as described above are present, yet in specific cases variations from the trends can be expected.

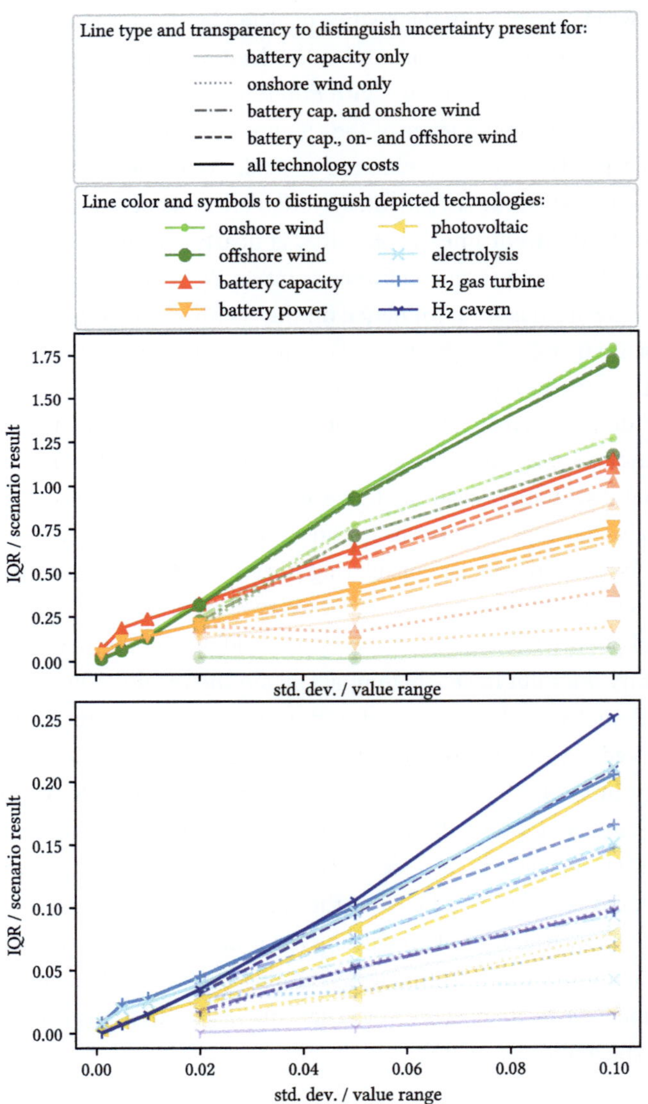

Figure 4.4: Influence of distribution width and number of uncertain parameters
on optimally installed capacities.

4.4 Summary of Systematic Errors if Uncertainty is Ignored

The considerations show that an overestimation of the expected optimization objective, i.e. the mean of the stochastic optimization, results if the input parameter variations are not taken into account. This is due to the mathematical properties of convex and thus concave functions, summarized by Jensen's inequality. Therefore an ESO which optimizes for system cost or cost of energy will yield a higher cost in the case where only one "mean-scenario" is investigated, compared to the investigation of the scenario with the underlying uncertainties in stochastic optimization. The investigation with a simplified model energy system reveals that the overestimation of the optimization objective and generally also the width of other parameters' output distributions increase with broader input parameter distributions and with the number of uncertain parameters. The specific impact of the uncertainties on other output parameters, i.e. technology implementations, which are not the optimization objective, depends on the parameter itself and its interaction with the uncertain input parameters.

Chapter 5

Optimization Model of the German Energy System under Uncertainty

The results presented in chapter 4 underline that without the consideration of uncertainty the interpretation of ESO results can be misleading. In order to investigate this further, a model of the German energy system that bears more characteristics of the real world system is designed. The structure of this model is similar to other models used for ESO with total system cost as optimization objective, whose results are or have been used to advice or influence policy making. In this chapter model structure and relevant input data are presented in detail.

In order to give an introduction to the model, its characteristics are categorized according to the different dimensions elaborated in section 2.1.2. The underlying and arguably most important aspect of a model is its aim, as the suitability of a model depends its intentions for use. The aim of this model is to evaluate the difference in information gain and interpretability of an exploratory stochastic optimization with the scenario approach for an ESO model that is used to evaluate the cost optimal structure of a future, mainly renewable based energy system depending on the allowed carbon emissions. Therefore the aim contains the following aspects:

- Structural similarity to models which are used as a basis for policy decisions or are relevant to the public opinion and decision making process for the optimal structure of a future energy system using little or no fossil energy carriers.

- Comparison of the impact on information gained and possible differences in interpretation between a scenario and stochastic optimization.

- Cost optimal structure of the future energy system under the taken assumptions.

- Evaluation of uncertainty regarding the cost optimal future energy system structure.

- Evaluation of the possibility of a widespread use of stochastic optimization and to facilitate its use and interpretability.

Because of the above mentioned aims some of the model dimensions are chosen in order to achieve similarity to the most widespread models. The model is an energy supply model focused on optimally matching supply and demand according to the model purpose definition from [34]. Its focus lies on the energy system, including electricity, heating and transportation sector and does not specifically take into account e.g. economy or demography. The model is a bottom-up linear programming optimization model that contains details regarding technology costs and conversion efficiencies. The geographical coverage is that of Germany, no links to the surrounding countries are included and it is assumed that there are not preexisting energy conversion technologies that can be used without investment cost. Regarding the time horizon the model can be classified as a long term model as the focus is the optimal energy system structure under technology cost conditions in the year

2050. A single year is optimized based on weather and consumption data in hourly resolution from the year 2018.

The model's data requirements can be divided into three categories, which are the basis of the following subsections' structure:

1. Demand data for heat, electricity and transportation considering temperature levels and type of transportation, details in section 5.1 on the next page.

2. Current cost and efficiency data of the different generation and conversion technologies, details in section 5.2 on page 73.

3. Probability distributions regarding probable future cost reductions, details in section 5.3 on page 95.

There are a range of implicit and explicit assumptions, some of which are included in the description of the model dimensions above. Additional important assumptions are made regarding the economic aspects. The annuity method is used, assuming the service life of a system or component as its depreciation time. For optimization a government funded investment is assumed, leading to the application of lending costs that can be expected for the German government. The interest rate is calculated as the real interest rate based on the Fisher theorem [119] and an assumption of 2 % inflation, which is in accordance with the inflation goal of the European Central Bank [120], as well as an assumption of 0 % interest rate, which is slightly above the medium interest rate of -0.38 % for 10 year German federal bonds between May 2019 and May 2020 [121]. The resulting real interest rate is -2 % a^{-1}.

The structure of the model is depicted in fig. 5.1. It focuses on electricity, transport and households as well as businesses, trade and services heating sector, in total taking into account about 94% of German final energy consumption. The remaining 6% are contributed mainly by air traffic which is not taken into account. The areas electricity, heat, H_2 and CH_4 represent forms of energy, the transformation between which is possible via transformation processes respectively technologies, represented by the labeled arrows. The demands to be satisfied are grid consumption on the transmission level of the German electricity system, electricity for an electrified transport sector and heat consumption in households, as well as businesses, trade and services. Storage options are available for each form of energy with batteries and flow batteries for electricity, caverns for the gaseous energy carriers and tanks for on site hydrogen storage as well as hot water storage for heat on a district heating or business level and household level respectively. For medium and high temperature heat molten salt respectively liquid metal storage is included in the model. Energy is introduced into the system by either renewable options, including on- and offshore wind power plants and photovoltaic plants, or by imported, fossil natural gas. In the following subsections the derivation of firstly the demand time series and secondly technology costs as well as efficiencies is explained in detail. The derivation of probability distributions and overview on the performed optimizations is given in the end of this chapter.

5.1 Energy Demand

Electricity, transport and heating are the three sectors of energy use considered herein. In the following, the approach how time resolved consumptions are derived is explained in detail for each of these three sectors. One important underlying assumption is that the demand for a certain service will remain the same in 2050 as it has been 2018. This assumption is obviously wrong in the sense that there will be changes, but may be the best guess, as influences for both, increasing and decreasing demand, are present in all three sectors and it is not possible to foresee any developments with a certainty that would, to the mind of the author, justify assuming a change in demand. Some of

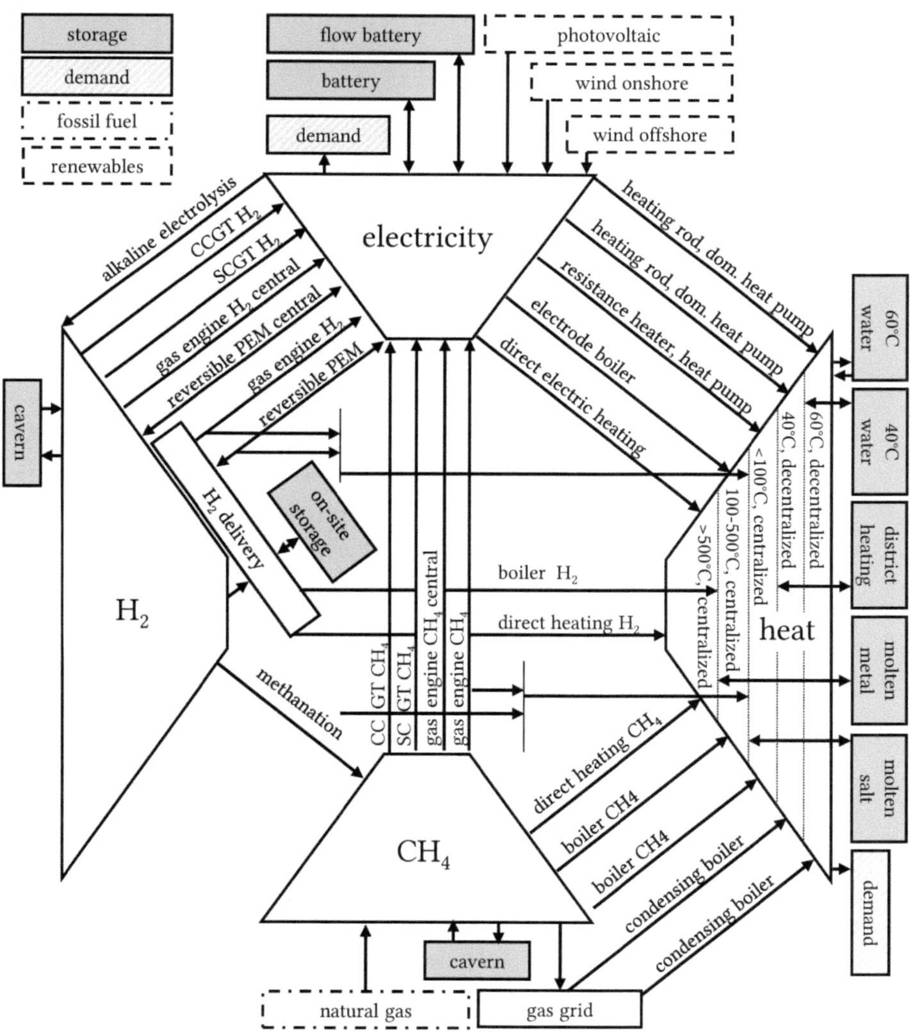

Figure 5.1: Structure of the ESO model and considered technologies. Abbreviations: domestic (dom.) combined cycle and simple cycle gast turbine (CCGT and SCGT), proton exchange membrane (PEM).

the counteracting influences on each of the sectors are mentioned in the corresponding sections.

5.1.1 Electricity

In order to obtain a time series for the electricity consumption in Germany, data from the Homepage SMARD.de [122] from the German federal network agency "Bundesnetzagentur" is used. The electricity consumption time series from 2018 for Germany available for download at [122] results when subtracting electricity export to neighboring countries and electricity consumption for filling pumped hydro storages from the sum of net electricity supply by the countries power plants to the grid and imported electricity [122]. As this data refers to net supply to the grid, all grid losses from transmission level to consumption are already included in the demand and are not be considered separately. This results in 498.8 TWh electricity use, representing 97% of the consumption accounted for in the official publication of the Federal Ministry for Economic Affairs and Energy [3]. Regarding the constant consumption assumption from 2018 to 2050 developments leading to increased electricity demand could be economic growth, a growth in population due to immigration [123] and growing electricity demand in datacenters [124, 125]. Trends such as electrification in transport and heating are taken into account with transport and heating demand and are considered in the respective chapters. There are also trends that seem to suggest a decrease in electricity demand, such as more efficient use of energy [126], the decoupling of gross domestic product (GDP) from energy consumption [127] as well as the possibility of a decreasing population in Germany because of low fertility rates [128]. There are further influences in either direction. To the mind of the author a reliable estimate that justifies the additional effort of determining the possible influence of these parameters is unlikely to be achieved.

5.1.2 Transport

To evaluate the transport sectors impact on the energy system, a time series for energy demand due to transport is derived. A method developed by the author that has been used for an energy system study on the state level of Bavaria, Germany [129] is applied. The methodology is visualized in fig. 5.2. It is based on traffic counting stations which are installed on federal roads "Bundesstraße" and the German Highway system "Autobahn". Vehicle occurrences in hourly intervals classified by vehicle type and type of street are published by the responsible federal agency [130]. The data from the available 1343 traffic counting stations is aggregated to four hourly time series by vehicle type, namely car-like, delivery, motorcycle and truck-like. The occurrences of each vehicle type are multiplied with the typical distance specific energy consumption of the respective vehicle type. This leads to a time series for each vehicle that is in proportion to its energy consumption if compared to the other vehicles time series. The sum of all the vehicle type specific time series is scaled to the sum of energy consumed in the German transport sector. This yields the current energy demand time series for the German transport sector in energy use per hour by vehicle type for each hour of the year for the currently applied transportation technology.

For a future energy system, in this case in Germany in the year 2050, a different structure of the transport sector can be expected. Especially private transport is a highly emotional topic and not the result of rational economic decisions [131]. Determining the future energy system by optimization between different types of possible propulsion and storage concepts, e.g. hydrogen, electric and synthetic fuels, seems thus unreasonable. Therefore, electrification is assumed as it is the most energy efficient approach [132]. According to the judgment of the author electrified transport seems a likely candidate for a dominant role in the future transport system because of current registration number [133] and charging power developments [134–136]. Therefore an efficiency gain is implemented for each vehicle type. The consideration of this

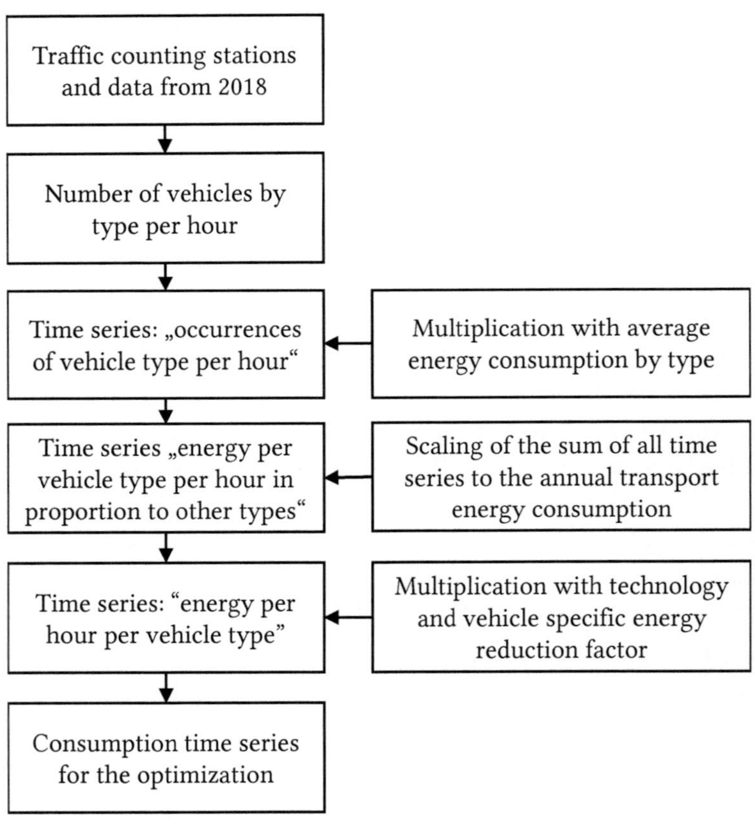

Figure 5.2: Procedure of time series generation for the energy consumption in the German transport sector according to [129].

effect is attained via a multiplication of each vehicle type's current energy consumption time series with a reduction factor, e.g. 0.6. This would represent a 40 % reduction in energy use for the same transport service supplied by the respective vehicle type. The sum of the resulting time series' represents the energy consumption in the year 2050 in the transport sector under the assumption that mobility demand remains the same and only the way of energy conversion to cover the distance changes.

In order to use the time series derived from the process outlined above it is important to consider which underlying assumptions are made and to understand the limitation of the resulting time series. The most important aspects as well as the numerical values used herein are summarized in Table 5.1.

Other forms of transport

The first simplification of the above methodology is that mainly road transport is taken into account. The actual German transport system is comprised of a variety of different means of transportation. Table 5.1 summarizes the different means of transport, their respective energy share and how they are taken into account in the energy system model herein.

Of the transport sectors final energy consumption of 2755 PJ, respectively 765.3 TWh, about 1.5 % are already included in the electricity sector due to electrification, mainly of railroad services [12, 13]. 15.4 % of transport energy demand, caused by aviation, are not taken into account. Therefore 84.6 % of the transport sector, corresponding to 647.43 TWh, are considered energy wise, 635.96 TWh of which in the transport sector time series, the rest via the electricity time series. Of the 647.43 TWh, 97.5 % are considered in form of time resolved data. The resulting traffic time series is scaled in order to contain the remaining 2.5 % not available in time resolved data, which are caused by city buses, water transport and diesel powered trains. The assumption is therefore that energy consumption of these kinds of transport correlates with the rest

Table 5.1: Different means of transport in Germany.

Transport	Energy share[a]	Time series considered	Energy use considered	Reference
Road	82.1 %	yes	yes	[137]
Aviation	15.4 %	no	no	[137]
Shipping				
Inland Waterways	0.4 %	no	yes	[137]
Railroad	2 %			[137]
Electrified	~1.5 %	yes[b]	yes[b]	[138]
Diesel Powered	~0.5 %	no	yes	[137, 138]
Public Transport				
City Bus	~1.2 %[c]	no	yes	[137, 139, 140]
Metro		yes[b]	yes[b]	
Tram		yes[b]	yes[b]	

[a] Share of final energy consumption of the inner German transport sector
[b] Taken into account with the electricity time series
[c] Part of road transport

of the transport sectors energy consumption. The assumptions should suffice, especially considering the great uncertainties in transport sector development until 2050. The main shortcoming of the chosen approach is the neglect of aviation. The production of renewable jet fuels in regions of the world with abundance of surface area [141–143] and a possible electrification of short flights [144] seem possible. In this case renewable generation capacity for the drop in fuel could be imported like crude oil and its products nowadays and only electricity for e.g. inter European flights parting from Germany will have to be generated, adding relatively little to the overall electricity generation.

Traffic counting stations

An additional source of inaccuracy is caused by the assumption that traffic counts can be converted to energy as described above. The distribution of both, highway and federal road stations is assumed to be representative of the complete road system in Germany. This can cause time lags between generated energy consumption time series due to recorded traffic if e.g. counting stations are mostly positioned outside of cities and traffic departs mostly from these. Yet this time lag can be considered smaller than an hour and therefore is neglected due to the herein chosen hourly time resolution. Other effects, resulting from the distribution of traffic counting stations, e.g. especially along roads relevant for local authorities or similar, cannot be ruled out.

Vehicle efficiency

Distance specific energy consumption of a moving vehicle is dependent on its velocity. Usually the velocity on federal roads and highways will not be equal. The resulting difference in consumption is not taken into account, as the number of counting stations on highways in proportion to those on federal roads, is not necessarily representative to the proportion of highways to federal roads. The procedure outlined in fig. 5.2 needs assumptions regarding energy consumption according to vehicle type and regarding achievable energy efficiency improvements. Table 5.2 depicts the vehicle types distinguished by the counting stations and how they are grouped herein. The current specific energy consumption is extracted from an analysis using a web interface [145] that allows drivers to record their measured consumptions by vehicle. Where sufficient data is available this database is also used for electric vehicles. As in the case of "car-like" and "truck-like" vehicles there is no sufficient data, the test results from [146] that include charging losses are used for passenger vehicles whereas delivery and "truck-like" vehicles are based on manufacturer data [147–153].

Table 5.2: Different means of road transport and energy efficiency ratio defined as consumption before and after electrification based on consumption data from the before mentioned references and the ratio of diesel to gasoline cars from [154].

Vehicle group	Vehicle type [130]	eff. ratio
car-like	car*	0.3
delivery	delivery vehicles	0.43
motorcycle	motorcycles	0.19
truck-like	special vehicles, truck*, semitruck , busses	0.523

* with and without trailer

Grid services from electric vehicles

The possibility of load shifting is many times mentioned as a potential advantage or necessity [134, 155] for an electrification of the transport sector. Herein, instead of assuming grid services from the vehicle, a "worst case" scenario is assumed. In this scenario vehicles will not supply grid services on a large scale. Instead, to the mind of the author, it seems reasonable to assume charging behavior similar to refill practices regarding liquid fuel powered cars, considering recent developments in high speed charging technology [134–136]. In addition, connecting high power charging stations that can serve many clients to the medium voltage grid could be more economic on a macroeconomic scale than empowering the low voltage grid to sustain home chargers that have to be purchased and installed for all vehicles. An other underlying assumption of the derived time series is therefore, that charging respectively consumption from the grid takes place when vehicles are moved, similar to present day refilling patterns.

5.1.3 Heating demand

Final energy consumption for thermal purposes including space heating, hot water, process heat, space cooling and process cold in Germany contributes about 1395 TWh to Germany's final energy consumption [3]. Table 5.3 depicts the structure of heat and cold use in Germany and how it is considered in the model. As in case of transport a part of the consumption is already accounted for by the electricity time series. The distinction in district heating, respectively centralized heating and decentralized heating is made, because different generation options for these types exists, mainly determined by the size of the generation unit. To take this into account the time series for each demand is divided according to the proportions of demand shown in table 5.3 in a time series for bigger generation units, centralized, and one for smaller generation units, decentralized. For these generation units prices and efficiencies can be different.

Table 5.3: Structure of heating Demand in Germany [3].

Type	District heating in TWh	Electricity in TWh	Other in TWh	Other accounted for
space heating	61.70	15.22	562.44	as decentralized
hot water	5.83	22.77	96.57	as decentralized
process heat	43.74	85.96	435.68	temp. dependent
air conditioning	-	9.58	1.42	neglected
process cooling	-	53.6	-	-

For each of the three generation demand types space heating, hot water and process heat, a time series is derived. Air conditioning powered by electricity and process cooling time dependencies are already considered in the electricity time series. Neglecting 1.42 TWh for air conditioning, in total 99.9 % of heat demand are accounted for in the model. In the following the generation of the necessary time series is explained.

Hot Water and Space Heating

Hot water and space heating demand are derived together, based on hourly consumption factors from Hellwig [156]. In order to assign building type and age specific energy consumptions, the distribution of total energy demand for heating and hot water is derived based on data from [157]. In order to allocate hot water demand to the different buildings, it is assumed that hot water demand is proportional to the number of residential units and independent of building age and type. With this assumption the derivation of the distribution of total demand depending on building type and age is possible, resulting in the distribution depicted in tables 5.4 and 5.5.

Table 5.4: Distribution of heating and hot water demand depending on building types in Germany based on data from [157].

Building Type	before 1979	after 1979
detached	0.41	0.22
apartment	0.27	0.10

Table 5.5: Distribution of heating demand depending on building types in Germany based on data from [157].

Building Type	before 1979	after 1979
detached	0.43	0.24
apartment	0.25	0.08

Hellwig calculates the relative daily demand of space heating based on a logistic growth function, extended by a constant value to take into account hot water demand. This function is used to calculate a days heat demand relative to the yearly demand based on the average temperature. In this work data from 506 weather stations in Germany is taken into account and the calculation is performed for each of the temperature profiles. Once the daily demand is determined, hourly demand profiles from [156] are applied and the time series for each building type and age are calculated once for space heating demand and once for the combined space heating and hot water demand. The resulting time series are weighted according to the structure depicted in table 5.4 and table 5.5, resulting in a single time series for each, space heating and combined space heating and hot water demand. Subtraction of one from the other yields the hot water demand time series. To achieve time series for central and decentralized generation the energies recorded in table 5.3 are assigned to the time series accordingly. This process to achieve the four time series, one for hot water demand supplied by centralized generation, one for hot water demand supplied from decentralized generation as well as one for space heating demand supplied via centralized generation and one for space heating demand satisfied with decentralized generation, is visualized in fig. 5.3. In a subsequent step, as well displayed in fig. 5.3, the centralized and decentralized generation is added. This yields three time series, centralized, decentralized and electric heat generation, for the two building types. Under the assumption that centralized generation does not change for older or newer buildings, the centralized generation for both building types is added. Since the electricity demand is already accounted for in the electricity time series, this yields three time series which are included in the model. These are decentralized, or local generation for old and new buildings, as well as centralized heat generation. Decentralized generation is taken into account divided by building type in order to be able to provide adequate heat generation efficiencies, especially for heat pumps, in the case of old and new buildings.

Process Heat Demand

Process heat demand is derived based on relative heat demand from [158] that distinguishes day and night consumption for each month and 43 different industry types. The procedure of time series derivation is depicted in fig. 5.4. In [158] demand is distinguished depending on three company sizes, for simplicity it is assumed that energy use in small, medium and big companies

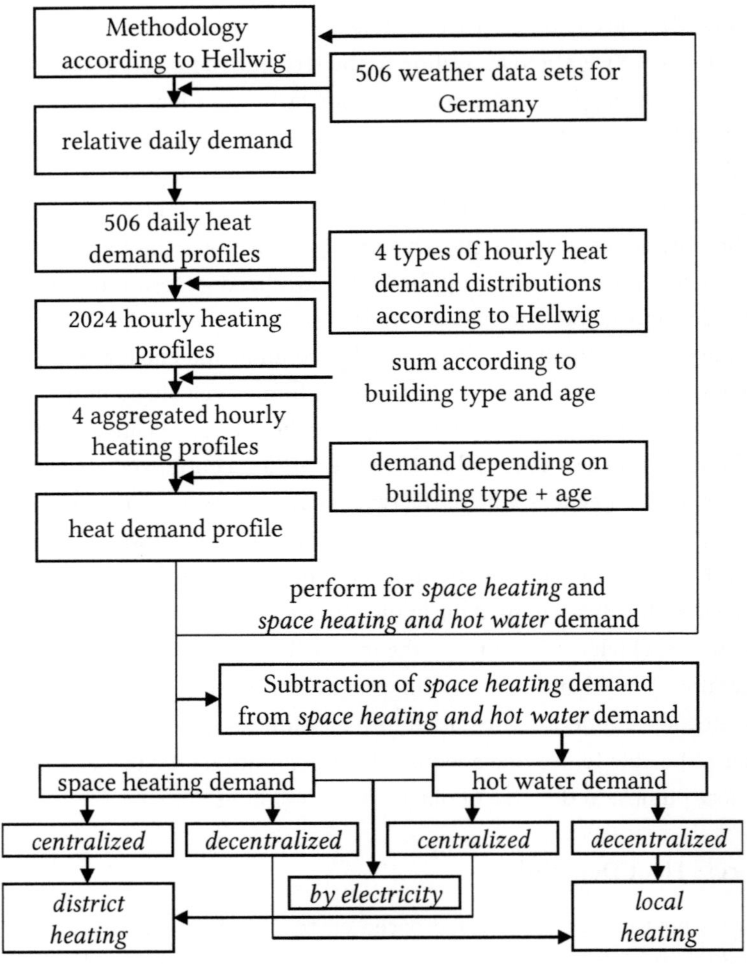

Figure 5.3: Procedure to derive heating and hot water demand time series.

relative to one another is 1:1:1. Each of the 43 industry types is assigned to one of the five biggest sectors according to the sectors in [159] resulting in six aggregated relative heat demands for five sectors that represent 92 % of heat demand [159] and one of 8 % that represents other industries. The aggregated relative load profiles are normalized and then scaled by their share of energy depending on the sector and temperature level according to [159]. The resulting normalized profiles are scaled such, that the sum of the hourly load profiles equals the yearly process heat demand in Germany depicted in table 5.3 according to [3]. This yields three time series for process heating demand according to the three temperature levels. For process heat demand no distinction is made between district heating and local generation, as it is assumed that the generation units are of sufficient scale to result in similar equipment prices for both cases.

5.2 Generation and Transformation

In this section, the model aspects regarding electricity and heat generation are explained. This concerns on the one hand the intermittent nature of power generation from solar and wind and on the other hand the techno-economic aspects including cost per capacity and efficiency values. In this context also the availability of a power plant or technological device is an important parameter. In order to achieve the calculated service provided by a plant with e.g. an availability of 0.8, i.e. 80 % of the time it is available, actually 1.25 times the installed capacity is necessary in order to satisfy demand. That means the actual installation and fixed annual costs for that service are 25 % higher. Availabilities are taken into account if available from literature, technology costs are adapted accordingly in the input parameter set, in the example case costs are assumed to be 25 % higher. Where no literature values or assumptions are stated an availability of 100 % is assumed.

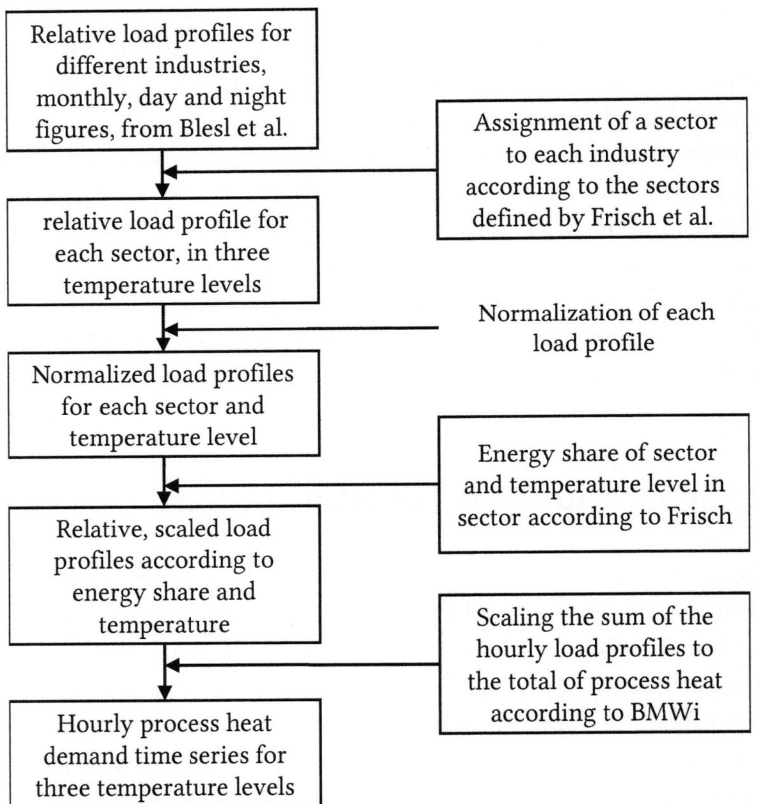

Figure 5.4: Procedure to derive process heating demand time series for the temperature levels low temperature (LT) <100 °C, medium temperature (MT) 100-500 °C and high temperature (HT) >500 °C. Input from BMWi, Blesl and Frisch [3, 158, 159].

The technologies considered in the optimization and visualized in fig. 5.1 have to be characterized regarding their economic aspects. This includes capital expenditure (CAPEX), characterizing the up front investment cost, and operational expenditure (OPEX) which consists of fixed operation and maintenance cost (FOM) as well as variable operation and maintenance cost (VOM). Also for transformation processes such as heat pumps, gas turbines and boilers, conversion efficiencies must be determined. In the following, each technology is characterized regarding present day investment and operational cost, as well as expected service life and efficiency values if applicable. Regarding economic parameters the model requires annual costs for each technology as an input. With the option to define additional fixed and variable generation cost this is either achieved by applying the annuity method to transform the investment cost to a fixed annual rate over the technology service life or by directly deducing the yearly cost from available data. The improvements regarding costs and efficiencies expected until 2050 are addressed in section 5.3.

5.2.1 Fluctuating Renewable Electricity Generation

Renewable electricity generation depends on weather conditions and cannot be scheduled like conventional electricity generation. In ESO renewable generation is usually represented by time series, which consist of the power output per installed peak power of the considered technology.
Herein time series derived directly from renewable generation in the past are utilized. This results in a time series that does not represent a typical year, but instead the specific year the empirical data is taken from. This has the disadvantage that data is not representative of an average year. Yet also in the case of typical data every real year will have a different generation profile. The disadvantage of typical year time series for onshore and offshore wind generation, solar generation, transport, electricity and heat consumption if they are created independently of one-another is, that they do not take into account

the interactions between these sectors. Using real past data in contrast has the advantage that interactions between weather and consumption behavior regarding e.g. lighting, heating, holidays, sport events, use of transportation services and many more, are automatically taken into account correctly. While the system, resulting from consumption behavior and renewable production time series, is not representative of a typical year, it is internally consistent. Therefore, this approach is used herein.

Renewable generation time series of onshore, offshore and solar power plants in Germany are available for download at [160]. Peak power specific generation time series are achieved by dividing generation time series by installed capacity values. In order to take into account that new capacity is constantly commissioned the evolution of installed capacity over time needs to be taken into account. To this end installed power in the end of each month is taken from [161] for onshore wind and solar power. In the case of offshore power information is missing at [161] for the year 2018, therefore installed capacity in the end of 2017 is taken from [161] and capacity additions by date are taken from [162]. Daily capacity values for onshore wind and solar power plants are achieved by linear interpolation from one month to the next, for hourly values constant capacities during each day are assumed. This results in peak power specific electricity production profiles for these three technologies that take into account the currently installed generation mix regarding age, location and plant type. Possible changes in plant design, e.g. lower cut in wind speeds for wind power plants or different angles for future solar power plants are not accounted for.

5.2.2 Renewable Generation

As technological aspects regarding conversion efficiency are already captured in the time series of renewable generation technologies, only the economic parameters and expected service life need to be determined. The information about current cost structures of renewable generation devices is not freely available and accessible estimates are usually outdated due to the fast evolution of generation cost in this sector. In order to estimate the current cost of photovoltaic, land based and offshore wind electricity generation, a hybrid approach is used. Benchmark levelized cost of energy (LCOE) from Bloomberg New Energy Finance [163] are compared to bidding results for photovoltaic and land based wind generation in Germany. Offshore wind bidding results are not included, as these have had fixed pricing schemes in the past. The only bidding results available are for plants due to be completed after 2021 and gains of electricity sales comprise part of the expected returns, which is why "zero subsidy" projects resulted in some of the auctions [164]. Therefore calculated LCOE is compared with bidding results, that represent the cost the generated electricity bears for the public. Both measures are comparable, as the LCOE includes the investors profit in the form of interest rates for their invested equity. Grid connection costs which are payed for by the public are not taken into account.

Figure 5.5 depicts the price development since the introduction of the bidding scheme for photovoltaic and land based wind generation in Germany. The shaded areas represent respective minimal and maximal successful bids, while the symbols represent the capacity weighted mean of the respective auction. All costs are given as reported, the figures are not adjusted by inflation. As significant yearly cost declines are reported for the considered technologies [165–167], the rise in auction results starting in 2018 seems counter-intuitive. Besides the role of financing costs, which may vary in economic cycles, the cost on a national level is also depending strongly on national legislation. Comparison of the auction results with the LCOE data from Bloomberg New Energy Finance [163], show a difference, especially regarding land based wind. Here the auction results in Germany after 2017 are 44 % higher than the calculated LCOE. Since the auction prices for Germany take into account the specific circumstances in the country, 2019 average auction results are used as a basis for the model. For offshore wind it is assumed that the underestimation of German wind cost shown by the land based wind data holds true, increasing the LCOE provided by [163] by 44 %. Resulting LCOE and full load hours (flh) derived from the time-series of fluctuating generation in this chapter are summarized in table 5.6.

Table 5.6: Assumed 2018 cost of renewable energy generation technologies, full load hours calculated based on the time series according to section 5.2.

Technology	LCOE € MWh^{-1}	flh h a^{-1}
Offshore Wind	102.2	3329
Onshore Wind	61.6	1686
Photovoltaic	55.1	911

Regarding the maximal installable renewable generation capacity there are different approaches to determining the upper limit of capacity that can be installed. Especially regarding wind power besides wind conditions, which are strongly dependent on the location, there are also limitations for the construction of wind turbines close to residential and military areas, as well as airports and natural reserves [168]. In this work the upper limit in the case of wind power is assumed to be 200 GW_{peak} for land based wind and 85 GW_{peak} for offshore facilities which is in line with the estimation of 198 GW_{peak} at 2 % land use by Bofinger et al. [169]. For offshore wind 85 GW_{peak} are assumed, in accordance with Fraunhofer ISE assumptions [170]. Photovoltaic installations are not limited by urbanized areas as rooftop solar is a viable option for these. In a recent publication from Fraunhofer ISE different sources calculating the potential for PV in Germany are cited and potentials calculated [171]. These are summarized in table 5.7.

No upper limit for photovoltaic installation is implemented since the potentials in table 5.7 sum up to almost 3 TW_{p}. At a yield of 900 $\text{kWh kW}_{\text{p}}^{-1} \text{ a}^{-1}$ this would produce over 2700 TWh a^{-1}, respectively about 115 % of the total

Table 5.7: Potential of photovoltaic installation in Germany based on [171].

Area	Potential GW$_\mathrm{p}$	Reference
Non-restriction open spaces	226	[172]
Energy crop replacement	1700	[171]
Building envelopes	800	[171]
25 % of future lakes over lignite mines	55	[171]
sealed settlement areas	134	[173]

energy demand accounted for in the model. The potentials listed in table 5.7 must be kept in mind when evaluating optimization results.

5.2.3 Gas to Electricity Processes

In the model two different gaseous energy carriers, H_2 and CH_4, are considered. The technologies that use these gases in order to generate electricity are similar. For gas turbines and gas engines the technology has first been developed for natural gas and later been adopted for hydrogen applications. In the case of hydrogen turbines until now there are no specialized hydrogen turbines in the scale of natural gas turbines, but there are several research and pilot projects under way which will enable either the adoption of existing power plants to hydrogen combustion or the design of hydrogen gas turbine power plants [176]. As for natural gas technologies more data is available from literature, these constitute the basis of the cost and efficiency analysis. Hydrogen gas turbines are evaluated based on their natural gas counterparts, results from [177] suggest that efficiency and power output can be comparable for hydrogen and natural gas turbines, therefore equal costs and efficiencies are assumed. Cost and efficiency data is summarized in table 5.8. Data for combined cycle gas turbines and simple cycle gas turbines is based on the average data of the five biggest plants in the respective category from [178]. To achieve fully loaded plant cost from bare bone costs, i.e. costs without additional

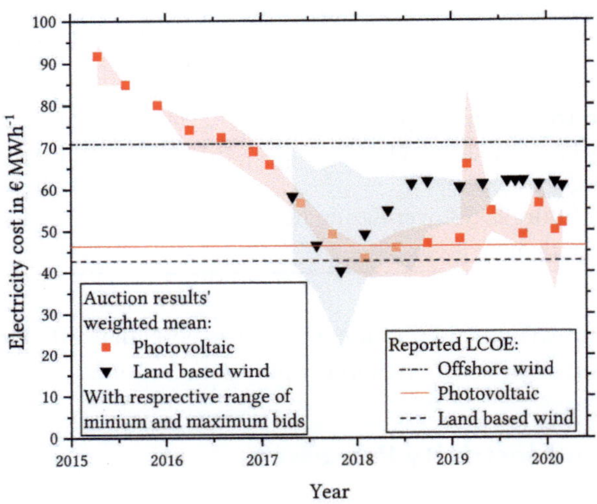

Figure 5.5: Comparison of auction results in Germany [174, 175] with reported LCOE [163].

equipment, additional costs of 30% (CCGT) and 100 % (SCGT) are included [178]. Because of the insights form [177] H_2 performance is assumed equal to that of natural gas in the case of CCGT and SCGT. Data on fixed and variable costs for gas turbine operation is not available directly from the contractors. For this reason data from four different reports [179–182] is used in order to determine a realistic estimate for large SC and CC gas turbine plants fixed as well as variable costs. As the goal is to model a national energy system and fixed and variable costs tend to decrease for larger systems the lowest cost given in the mentioned reports is used. In the case of variable cost for CCGTs the mean of the acquired data points is used, as no trend is discernible. The data depicted in table 5.8 is not adjusted for inflation, as cost savings and inflation are expected to cancel one another [182].

Cost data for co-generation is depicted in table 5.9. Performance and cost parameters for gas engines are taken from [183] for a 10 MW machine in the case of performance data, respectively 8.5 MW for cost data, as it presents the upper limit of the data range. Gas engine manufacturer 2G Energy AG offers a gas engine designed for natural gas and retrofitted for hydrogen combustion, the agenitor 412. On its homepage the company provides performance data for both fuels [184, 185]. The comparison reveals a 20 % power rating decrease for the hydrogen machine, resulting in 25 % higher specific cost of the hydrogen engine, if total cost is assumed to be equal. As variable and fix cost for the engine are assumed to stay constant after retrofitting, a 25 % cost increase is taken into account for these power, respectively energy specific cost components as well. Fix cost is assumed as 4.5 % of initial investment [186]. Initial investment is calculated by adding additional costs for transportation, installation etc. of in total 62 % to the engine cost [183]. Efficiency for the case of hydrogen powered gas engines is assumed to be equal to that of natural gas powered engines despite data from 2G Energy AG showing a difference of 0.8 % points, favoring natural gas [184, 185]. That difference is assumed to decrease if an engine is designed specifically for hydrogen and with increasing engine size. Yet it is possible that lower volumetric energy density may still result in lower efficiencies due to more losses because of increased friction losses in the bigger engine needed for hydrogen combustion.

Proton exchange membrane electrolysis cells (PEM EC) are taken into account for use with hydrogen only, methane conversion to hydrogen is not considered. Reported investment cost is based on the analysis for a 5 MW_{el} system in [187]. The FOM, 2 % of OPEX, includes stack replacement cost according to [188].

Lifetime for gas turbine systems is assumed to be 30 years, availability for SCGT is taken as 93.5 % and for CCGT at 89 % based on [191]. Thermal efficiency of gas engines is calculated by assuming a total system efficiency

Table 5.8: Cost and efficiency data of gas turbines. Abbreviations: lower heating value (lhv), efficiency η.

	Fuel	CAPEX € kW^{-1} [178]	VOM € MWh^{-1} [179–182]	FOM € kW^{-1} a^{-1} [179–182]	η % lhv [178]
SCGT	CH$_4$	382.6	7.3	2.4	41.9
	H$_2$	382.6	7.3	2.4	41.9
CCGT	CH$_4$	741.7	2.5	6.1	61.8
	H$_2$	741.7	2.5	6.1	61.8

Table 5.9: Cost and efficiency data heat and electricity co-generation.

	Fuel	OPEX € kW$_{el}$$^{-1}$ [183–185, 187]	VOM € MWh^{-1}	FOM € kW^{-1} a^{-1} [186, 188]	η % lhv [183, 189, 190]
Gas Engine	CH$_4$	651.9	4.4	29.3	46.9[a], 42.1[b]
	H$_2$	814.9	4.4	36.7	46.9[a], 42.1[b]
PEM FC Mode	H$_2$	1200	0	24	30[a,c], 50[b,c]
PEM EC Mode	H$_2$				62.4[a], 10[b,c]

[a] electrical efficiency
[b] thermal efficiency
[c] rounded, based on [190]

of 89 %, which is in agreement with reported data [183]. Subtracting electrical efficiency yields heating efficiency. Lifetime is assumed to be 15 years including overhaul, which is already accounted for by variable cost [183]. Gas engine availability lies at 98 % according to [192]. PEM systems are assumed to degrade 1 % a^{-1} during a useful service life of fifteen years, which is inside the range given by [189], the given electrical efficiency in electrolysis cell mode is the resulting average efficiency during service life [189]. This way degradation is included in the model. Thermal efficiency is taken from [190] and rounded, as it is presented for a prototype system and deviations can be expected for industrial sized systems. Electrical and thermal efficiency assumptions are based on [190] as well. The availability of PEM is taken as 97 % based on [193]. For hydrogen based gas engines and PEM systems two different systems can be used, either with or without heat integration. PEM fuel cell (PEM FC) systems with heat integration are assumed to be in a decentralized location and therefore not close to hydrogen storage caverns. For these it is assumed that hydrogen delivery costs of 24.548 € MWh^{-1} apply at an efficiency of 98.5 % based on compressed gas delivery over a distance of 50 km at 15 t day^{-1} [194] . For the case of hydrogen production it is assumed that heat integrated systems produce hydrogen on site, which than has to be either stored on side or transported back to the cavern. If transported back to the cavern hydrogen delivery costs and energy losses apply as above. Stored on site a storage system has to be build, that currently costs about 15 \$ kWh^{-1} and is projected to have a minimum cost of 2.5 \$ kWh^{-1} as a lowest cost case [195, 196]. Based on this data 4.55 € kWh^{-1} are assumed for 2050 hydrogen on site storage without variable costs, at 1 % annul fixed costs, with a system life of 15 years and negligible losses.

5.2.4 Electricity to Gas

Besides PEM electrolysis hydrogen generation can also be achieved via alkaline electrolyzers (AEL). Model data for AEL is included in table 5.10. The given efficiency is the average efficiency of a system, starting with 71 % lhv efficiency and degrading 0.875 % a^{-1} as medium value according to [189] for a service life of 25 years [197]. This yields a medium efficiency of 49.125 % as recorded in table 5.10. Investment cost is based on the cost of a 5 MW_{el} system, that has been calculated based on literature reports of AEL projects [187], FOM is 2 % a^{-1} [188]. An availability of 97 % is assumed.

Table 5.10: Cost and efficiency data of alkaline electrolysis.

	CAPEX $€\,kW_{el}^{-1}$ [187]	FOM $€\,kW_{el}^{-1}\,a^{-1}$ [189]	η % [189, 197]
alkaline electrolysis	1100	22	49.125

Methanation

In order to take into account established natural gas technologies as well as the above mentioned emerging hydrogen alternatives, methane must be present in the system. One source of methane is the import of fossil natural gas, here simplified as pure methane. In order to allow for long term storage and the possible beneficial double use of equipment for fossil and synthetic gases, the process of methane production from hydrogen is introduced in the model. The process is called methanation, the predominantly discussed way of producing methane from hydrogen and carbon dioxide is chemical methanation [198]. Other alternatives, such as biological pathways [199, 200] are not considered herein. Cost and performance data is summarized in table 5.11.

Variable cost is not mentioned in the analyzed literature and therefore calculated based on the specific CO_2 consumption of the process. [201] give a

Table 5.11: Cost and efficiency data of the methanation process.

	CAPEX $€ \, kW_{hhv,CH_4}^{-1}$ [187]	CO_2 supply $€ \, MWh^{-1}$ [201, 202]	FOM $€ \, kW_{hhv,CH_4}^{-1} \, a^{-1}$ [203]	η % lhv [202]
Methanation	600[a]	2.4[a]	45	83[b], 10[c]

[a] based on reference with 50 € ton⁻¹, adapted with own calculations
[b] chemical efficiency
[c] thermal efficiency, assumed up to 500 °C

mean cost of 51 \$ ton^{-1} CO_2, that is simplified to 50 € ton^{-1} CO_2 by [202]. This value has been used herein to calculate variable cost. A thermal efficiency of 10 % is assumed as viable, an additional cost for the possibility to use heat is not assumed. The technical service life for the methanation reactor according to [203] is 20 years, avialability according to [204] cited by [188] is 85 %. Cited FOM is based on 7.5 % of CAPEX according to [203].

5.2.5 Gas to Heat

The produced hydrogen and fossil or via methanation generated methane can be used in the above mentioned processes to produce electricity or electricity and heat in co-generation applications. Another option is the direct use of gas for heating in furnaces and boilers. The technologies are listed in table 5.12 according to the temperature level they can supply and their application either in centralized, decentralized or process heating applications. It is assumed that hydrogen devices with cost and performance data comparable to natural gas devices will be available, therefore literature figures for natural gas applications are assumed for hydrogen combustion as well. For the transport of the respective gases the use of the existing natural gas grid for methane is assumed. As hydrogen cannot be transported with the existing grid without interventions on a technical level, it is assumed that hydrogen can only be used in central heating applications and not in decentralized, i.e. private heating applications. For the delivery of H_2 to the costumer, cost and efficiency are assumed as detailed on page 83. Methane delivery via the gas grid has negligible energy losses based on total gas losses in the gas grid [205] in comparison to total natural gas consumption in Germany [206]. Methane delivery cost varies between 42.66 € MWh^{-1} for applications smaller 7 MWh per year and 7.86 € MWh^{-1} for applications of more than 1.5 GWh per year [207]. For central heat generation a medium cost of 9.6 € MWh^{-1}, derived from the specific cost between 300 MWh a^{-1} and 1.5 GWh a^{-1}consumption, and the maximum of 42.66 € MWh^{-1} for decentralized applications are assumed. Power specific grid costs are neglected. Since changes to the technologies design are generally necessary in order to fire different fuels, no dual fuel ability is accounted for. Table 5.12 depicts the relevant parameters.

Table 5.12: Cost and efficiency data of gas fired heating applications.

	CAPEX € kW$_{heat}^{-1}$ [208, 209]	FOM € kW$_{heat}^{-1}$ a^{-1} [210]	Service life a [210]	η % lhv [210]
decentralized				
<100 °C:				
condensing boiler	200[a]	6	18	92
centralized				
<100 °C:				
boiler	90	3.15	25	95
100-500 °C:				
boiler	90	3.15	25	88
>500 °C:				
direct heating	90[b]	10.8	16	50

[a] installation cost of 1750 € per system of 19.2 kW is assumed and included
[b] assumption

5.2.6 Electricity to Heat

Regarding heat generation by electric means temperature as well as equipment size are important regarding performance and cost of the systems. Therefore the distinction in low temperature centralized and decentralized heating is made, as well as medium and high temperature heat generation. For medium and high temperature no distinction is made between centralized and decentralized use, as equipment size in industry use is assumed to be sufficient to allow for cost reduction to be similar to central units.

Decentralized heat is supplied mainly for the heating of housing and therefore in the lower temperature regime. The technologies based on electricity considered herein for that purpose are direct electric heating via heating rods and a heat pump with integrated storage for residential applications. Equation (5.1) is the fitting result based on performance data of an air source heat pump provided by Vaillant [211] with ΔT as temperature difference between ambient air and flow temperature.

$$COP = \begin{cases} 6 & \Delta T \leq 18 \\ -0.0063\Delta T^2 + 0.2307\Delta T + 3.905 & 18 < \Delta T < 46 \\ 1 & \Delta T \geq 46 \end{cases} \quad (5.1)$$

The calculation of the coefficient of performance (COP) is based on temperature difference between ambient temperature and flow temperature according to building type. Ambient temperature is taken into account for 506 measured hourly temperature time series in Germany. The COP is calculated based on eq. (5.1) and a flow temperature of 40 °C for "new buildings" and 60 °C for "old buildings" according to the specifications made in section 5.1, yielding an hourly COP time series coherent with the weather data of 2018. For both building ages an aggregated hourly COP time series is calculated by averaging the 506 distinct time series. Storage size is calculated based on the given volume of 300 liters and a return temperature of 30 °C for new and 40 °C for old houses respectively. The parameters characterizing low temperature electricity to heat technologies are listed in table 5.13. Variable cost is assumed to be zero for the considered technologies.

Table 5.13: Cost and efficiency data for low temperature decentralized heat supply via electricity.

	CAPEX € kW$_{heat}^{-1}$ [212] [213]	FOM € kW$_{heat}^{-1}$ a^{-1}	η - [211]
heating rod	10.5	0	0.99[e]
heat pump	703.8	8.3[d]	eq. (5.1)
storage[a]	capacity in kWh: 7[b], 3.5[c]		0.9 d^{-1}

[a] part of the heat pump system, included in heat pump cost
[b] old buildings
[c] new buildings
[d] 100€ unit^{-1} a^{-1} assumed
[e] own assumption

Installation cost of 100 € per unit for the heating rod and 1250 € per unit for the heat pump is assumed and included in the specific investment cost in table 5.13. As service life 15 years are assumed for both, heat pump and heating rod.

Centralized supply with low, medium and high temperature heat based on electricity in the model is realized with electric hot water and steam generation units, as well as a ground source heat pump for low temperature heat. Data for these technologies is summarized in table 5.14. For the heat pump 15 years and for all other technologies 20 years of service life are assumed.

5.2.7 Storage Applications

For the matching of fluctuating demand and volatile renewable generation, storage options are required in the energy system. In this model energy storage for electricity, heat and gaseous energy carriers are considered. If

Table 5.14: Cost and efficiency data of centralized heat supply via electricity.

References	CAPEX $€\,kW_{heat}^{-1}$ [214] [215]	FOM $€\,kW_{heat}^{-1}\,a^{-1}$ [214]	η - [214]
<100 °C			
heat pump	440	4.4	2.8
resistance heater	56	0.56	0.99[a]
100-500 °C			
electrode boiler	300	3	0.98[a]
>500 °C			
electric direct	350[b]	3.5[b]	0.98[a]

[a] own assumption
[b] own assumption based on [214]

round trip efficiencies (RTE) are given, it is assumed that the square root of the RTE applies for storing energy in the storage system as well as for the withdrawal of energy from the storage.

Electricity storage is accounted for by batteries, represented by lithium-ion batteries as an example of mature battery technology and redox flow batteries. Battery cost is divided in the cost for energy and power of the battery and sales tax is subtracted from the total system cost in order to calculated net cost. Data for differently sized systems provided by [216] is used to derive eq. (5.2) for the battery system cost c_{System} with P as power of the battery system in kW and W the storage capacity of the system in kWh.

$$c_{System} = AW + BP \tag{5.2}$$

Using the data provided by [216] and multiplying battery pack cost per kWh with a factor, adjusted in order to lead to a consistent formulation according to eq. (5.2) and assuming euro to dollar conversion as 1.1 yields

$$A = 0.356 c_{System} = 257.6\ €\,kWh^{-1} \tag{5.3}$$

with

$$c_{System} = 209\ \$$$

and the factor for the cost of power calculated to

$$B = 257.6\ €\ kW^{-1}. \tag{5.4}$$

Complete cost and performance data for battery storage is summarized in table 5.15.

Table 5.15: Cost and performance data of battery storage.

CAPEX		FOM	RTE	self-discharge
€ kW^{-1}	€ kWh^{-1}	€ kW^{-1} a^{-1}	-	% month^{-1}
[216]	[216]	[217]	[217]	[218]
247.9[a]	257.6[a]	33.64	0.9[b]	1

[a] extended with own calculations
[b] [217] give 0.85 as current value
herein 0.9 is assumed as 2050 RTE

According to a comparison of different battery cost reports [216] variable costs can be assumed to be zero, resulting in the relatively high fixed cost, that also includes degradation, by assuming medium instead of initial storage capacity. Therefore it is not necessary to take extra measures to include capacity loss in the model. Lifetime of the system is 15 years [216].

In the case of redox flow batteries cost data is difficult to derive from literature. Current system capacity costs range between 1680 $ kwh^{-1} for zinc-bromide redox flow batteries in 2016 [219] and 31.3 $ kwh^{-1} if the cost break down from [220] is applied to the system cost given in the current investors presentation from Cell Cube Energy Inc. [221]. In this range of possible system costs, the system cost based on a cost breakdown from [222] is used, as the authors also recognize the diverse costs stated in different publications. The methodology makes use of a life cycle cost model, a review of literature data and manufacturer information [222]. Fixed operating costs are set to 2 % of investment cost per year and system life is assumed to be 15 years in accordance with performance data based on various publications evaluated in [223]. Cost and performance data is summarized in table 5.16.

Table 5.16: Cost and performance data of redox flow battery storage.

CAPEX		FOM	RTE	self-discharge
€ kW^{-1}	€ kWh^{-1}	% of CAPEX a^{-1}	-	% a^{-1}
[222]	[222]	[223]	[221]	[224]
1470	248	2	0.75	1

The storage of energy carriers at a large scale is state of the art. Storage caverns for about 154.2 TWh, as well as pore storage facilities with a capacity of 85.9 TWh of natural gas were in use in Germany in 2019 [225]. This storage is assumed as existing and implemented in the model without extra cost. The efficiency of storage is assumed to be 97 %, accounting for leakage, compression and all other losses. Hydrogen cavern performance and cost data is taken from [226]. As the data provided by [226] explicitly excludes all surface facilities necessary for storage of hydrogen, an additional 50 € kW^{-1} are assumed for this. The impact of this assumption on the distribution of total storage cost between power and capacity cost is visualized in fig. 5.6. For both, caverns and the corresponding surface facilities, a service life of 35 years is assumed, based on [226]. The parameters used for the implementation of cavern storage in the model are given in table 5.17.

Specifications of heat storage applications are highly dependent on temperature level and system size. On the decentralized level a certain amount of storage is already included in the heat pump system considered according to table 5.13. Data for the different technologies is summarized in table 5.18. The medium and high temperature storage technologies molten salt and liquid metal are in different stages of technological readiness. Molton salt storage is regularly used in the context of concentrated solar power plants to either shift

Table 5.17: Cost and performance data of cavern storage.

	CAPEX		FOM	Capacity limit	η
	€ kW^{-1}	€ kWh^{-1}	% CAPEX a^{-1}	TWh	-
		[226]	[223]	[225, 226]	[226]
methane	-	-	-	238.2	0.98[a]
hydrogen	50[b]	0.143	2.5	1600	0.97

[a] own assumption

[b] own assumption, see fig. 5.6 for evaluation of impact

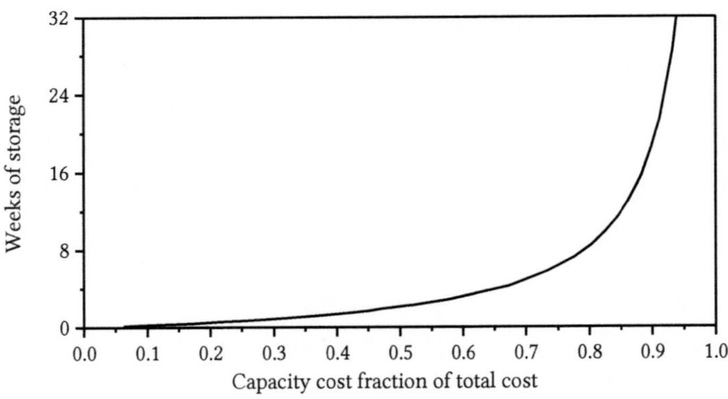

Figure 5.6: Impact of 50€ kW^{-1} surface facility cost assumption for hydrogen storage in caverns.

production completely in times without solar radiation in order to complement photovoltaic electricity production or to equalize output of the power plant and produce electricity after sunset [227–229]. Cost and performance data is reported by different sources including [219, 230, 231]. Regarding liquid metal storage there are no operating liquid metal storage sites to the knowledge of the author. Publication [232] evaluates the prospect of liquid metal as heat storage medium compared to molten salt systems and evaluates the system cost of liquid metal storage based on the difference in material costs and the necessary design changes. Comparing the specific system costs regarding only the elements necessary for heat storage and not taking into account those for the operation of the concentrated solar power plant, shows cost benefits for the liquid metal systems, partly due to lower storage material cost [232]. Since in this work the goal is to optimize a future energy system, it can be assumed that if liquid metal storage is economically more attractive than molten salt storage, it will be used for medium temperature applications as well. The proposed working temperature of the liquid metal storage is between 250 °C and 1200 to 1400 °C [232]. It must be assumed that a medium temperature storage makes use of the whole temperature range, minimizing the cost for energy storage, while a high temperature system would only be able to operate the upper end of that range. To take this into account, it is assumed that high temperature storage will cost twice as much as medium temperature storage, as it can use only half of the possible temperature range, e.g. 775 °C to 1300 °C instead of 250 °C to 1300 °C, while neglecting a possible temperature dependence in heat capacity. Losses and operating costs for the liquid metal system are assumed based on the reported numbers from molten salt storage.

Lifetime for the different systems is taken as 20 years for liquid salt and metal storage according to [231, 232]. For hot water storage systems on a centralized level 25 years are reported [234]. Decentralized low temperature hot water storage systems are assumed to last 25 years as well. Variable costs

Table 5.18: Cost and efficiency data of heat storage applications.

	CAPEX € kWh$_{heat}^{-1}$ [219, 230, 233, 234]	FOM € kWh$_{heat}^{-1}$ a^{-1} [234, 235]	self-dis. % h^{-1} [231]	η - [230, 231]
decentralized				
<100 °C:				
before 1974	37.7[b]	0.86[a]	0.298[b]	0.99[a]
after 1974	75.4[c]	1.72[a]	0.298[c]	0.99[a]
centralized				
<100 °C:				
heat accumulator	2.9	0.03	0.04[a]	0.99[a]
100-500 °C:				
molten salt	32.5	0.813[a]	0.127	0.95
>500 °C:				
liquid metal	performance like molten salt, twice its cost			

[a] own assumption
[b] for ΔT of 20K, linear scaling of losses
[c] for ΔT of 10K, linear scaling of losses

are assumed to be zero for all heat storage applications.

5.2.8 Fossil Energy Carriers

In order to allow for a fossil energy source in the energy system model, the use of natural gas is allowed. With the goal not to account for the trading gains of energy companies, exporters and importers, a price floor is chosen that should represent a low estimate of the cost to tap fossil natural gas reserves. With EU import prices from about 4.9 € MWh^{-1} in May 2020 [236] and cost estimates for current gas production from 3.35 € MWh^{-1} to 3.9 € MWh^{-1} and future production from 4.18 € MWh^{-1} to 7.65 € MWh^{-1} [237], a production cost of 5 € MWh^{-1} is assumed. To take into account the damage from carbon dioxide emissions of 240 € t^{-1} according to [238], 47.33 € MWh^{-1} are added at an specific CO_2 emission of 0.197 t MWh^{-1}, resulting in a total cost of 52.33 € MWh^{-1} for natural gas in this model. This is an intend to assume real costs, in the form of procurement and environmental costs, instead of prices for the commodity natural gas, as international markets are mainly dominated by shortage of supply and thus do not represent costs, but rather costumer values that are traded.

5.3 Probable Cost Reductions

In section 5.2 current technology cost for each of the models technologies is reported. Since the aim of this work is to optimize a future energy system in the year 2050, costs must be adapted to expected developments. A distinction is made between technologies with expected significant change in cost and other technologies, that are expected to have little or no change of costs. In the second group established technologies with moderate learning, respectively experience rates are included, such as gas turbines or hot water storage. It is important to note that this is a simplification as cost changes of these technologies will to a certain extend almost certainly occur until 2050. In the second group technologies with expected greater change in cost are considered, including technologies that currently display significant cost reductions on a year to year bases, such as photovoltaic and wind power plants or battery storage. Cost changes are also considered for technologies that are not yet employed on a great scale, but are promising in terms of cost reduction and contribution to the future energy system, such as PEMEC, methanation or redox flow batteries. Changes regarding system efficiencies, as considered in the simplified example in chapter 4, are not taken into account.

In order to address the uncertainty that expected cost reductions entail, data from literature regarding the range of expected minimal and maximal cost reduction in the future is used to define the uncertainty of the future system cost distribution. This investigation is mainly based on data provided by the National Institute of Renewable Energies in the context of the Annual Technology Baseline (ATB) [239]. The ATB contains current and future cost estimates for the most important renewable energy technologies including wind power, solar power as well as Li-Ion battery technology. The estimations are based on one or a combination of the following: expert elicitation, relative technology and cost improvements from recent literature and the application of experience or learning rates in combination with an energy system model to predict future capacity growth [239]. The future estimates are divided in a mid and low cost path, as well as a high or constant cost path until 2050. In the case of offshore wind generation it is stated that the mid assumption was asked from the experts to have 50 % probability of exceedance, while the low assumption should have a ten to thirty percent of exceedance. For other technologies no confidence levels of the intervals are stated. Since the probability distribution for a future technology can not be determined empirically, a normal distribution is assumed as explained in chapter 4. In order to define the distribution mean and standard deviation are assumed for each technology. The mid scenario is assumed as the mean and the difference between mid and low scenario as 1.96 standard deviations, corresponding to a five percent exceedance probability. This assures a uniform distribution is achieved even when the difference between high and mid scenario differs from that between mid and low scenario. The low scenario is chosen as a reference to determine the standard deviation, as a cost of zero is a natural boundary which should not be exceeded. Where no such data are available from the ATB standard deviations are based on the authors judgment. For all costs it has been assumed that the costs mentioned in section 5.2 are 2017 costs and cost reductions apply for the year 2050.

In order to take into account the correlation between the development of different technologies, cost breakdowns from the literature are applied, splitting up the total technology cost in component groups. Cost reductions are applied to the individual component groups in such a way, that the weighted sum of the component group cost reductions equals the technologies system cost reduction. If for individual component groups expected cost reductions are known from literature, these are either scaled to achieve component cost reduction, or the remaining components, for which no cost reductions are specified, are scaled in order to yield the defined overall cost reduction depending on whether there is information for all or only a part of the component groups cost reduction. Standard distributions are adjusted in a similar manner, where a components' standard distribution is not defined either by literature or by calculation from other technologies, the technologies overall standard deviation was taken as the components' standard deviation scaled by a factor in order to achieve the variance defined for the system cost distribution resulting from the weighted sum of all component group variances.

The technologies with expected significant cost change are listed in table 5.19 with data provided for expected mean cost changes and expected standard deviations. The component group breakdown where a correlation with other technologies component groups is expected is depicted in the corresponding tables listed in table 5.19. For photovoltaic and wind power applications the cost reduction is evaluated on a LCOE bases. For storage and conversion technologies a CAPEX reduction and equivalent OM reduction is assumed. In renewable electricity generation, especially in the case of wind power, LCOE cost reductions are expected due to system changes such as e.g. floating offshore wind turbines, that might lead to increased CAPEX which in turn will be overcompensated by higher full load production hours compared to current offshore wind turbines. The possible change of renewable production time series, e.g. a more even production over time, is not taken into account, only the LCOE impact of such a development is considered. This simplification must be kept in mind during result analysis, especially regarding storage applications and installed capacities.

Each technologies component groups and the respective cost reductions and standard deviations are depicted in tables B.1 to B.8 in appendix B. The consideration of the use of similar subsystems, roughly represented by the component groups, leads to a correlation between the technologies probability

Table 5.19: Technology cost reductions and distributions until 2050.

Technology	rel. cost mean 2050		rel. SD. 2050		details in
photovoltaics	53.10%	[239]	32.50%	[239]	table B.1
onshore wind	55.47%	[239]	21.69%	[239]	table B.2
offshore wind	27.07%	[239]	12.98%	[239]	table B.3
battery capacity	41.18%	[239]	26.12%	[239]	table B.4
battery power	40.87%	[239]	26.16%	[239]	table B.5
PEM	24.17%	[188]	20%	a.a.	table B.6
AEL	40.00%	[188]	20%	a.a.	table B.7
methanation	46.67%	[188]	20%	a.a.	
redox flow power	51.82%	[223]	20%	a.a.	table B.8
redox flow capacity	24.19%	[223]	20%	a.a.	
molten salt storage	29.41%	[219]*	20%	a.a.	
liquid metal storage	29.41%	a.a.	20%	a.a.	

min. values from literature assumed with 5 % exceedance probab. to calc. SD
a.a.: authors' assumption
*given 2030 cost reductions assumed as 2050 values

distributions. These correlations, calculated as Pearson correlation by dividing the covariance of the two compared distributions with the product of their standard deviations, are depicted in table 5.20. The most notable correlation is between on- and offshore wind due to the assumption that improvements in turbine and power electronics development have an influence on both technologies. Correlations not caused by shared subsystems but by chance are present, e.g. the maximum correlation by chance of -0.12 between molten salt or metal storage and battery capacity as well as between PEM and methanation. For methanation, redox flow battery capacity and molten storage no shared subsystems to any of the technologies have been taken into account.

Table 5.20: Pearson correlation coefficients for technology cost reduction distributions resulting from the use of similar subsystems.

		T2	T3	T4	T5	T6	T7	T8	T9	T10	T11
T1:	photovoltaic	0.01	-0.03	0.04	0.07	0.06	0.05	-0.07	0.08	0.06	0.13
T2:	onshore wind		0.66	0.01	0.13	0.05	-0.07	0.04	-0.01	-0.04	0.02
T3:	offshore wind			0.06	0.12	0.10	-0.07	0.04	0.02	-0.04	-0.04
T4:	battery capacity				0.00	-0.07	0.05	-0.01	0.06	0.08	-0.12
T5:	battery power					0.22	0.11	-0.01	0.01	0.06	-0.04
T6:	PEM						0.14	-0.12	0.07	-0.03	0.04
T7:	AEL							-0.03	0.09	0.05	0.03
T8:	methanation								-0.04	-0.02	0.00
T9:	redox power									0.01	-0.06
T10:	redox capacity										-0.01
T11:	molten storage										

5.4 Optimization Planning

The investigations performed herein are undertaken with the aim to achieve the goals outlined in chapter 3. This includes the following:

1. Investigation of the probable, optimal layout of a German energy system under island and greenfield assumptions.

2. Experimental investigation of the findings in chapter 4 concerning the influence of Jensen's inequality on optimization problems.

3. Comparison of scenario results, i.e. the result of a single optimization with the mean cost of each technology as cost assumption, with the stochastic optimization results.

4. Use of clustering and factor effect methodology in order to improve information gain and communicability of the stochastic results.

5. Investigation of the possibility to minimize computational effort for stochastic optimization with the use of less distributions consisting of fewer optimizations or a meta model.

The optimizations performed in order to investigate these aspects are depicted in fig. 5.7. The optimization procedure outlined therein is based on investigating the results of an ESO as it might be performed in a publication relevant for public opinion making and political decision making, if stochastic optimization was used. As recommended by Morgan et al. [52] probability distributions are applied for parametric uncertainty of empirical and chance parameters. For parameters influenced by the policy makers decision, decision variables according to Morgan et al., such as in this case the upper limit of on- and offshore wind power plants as well as the acceptable maximum carbon dioxide emissions, are assumed as fixed in the analysis. In the case of the carbon emission limits, the limit is varied in three policy relevant steps. Therefore the analysis is performed three times with imposed minimum carbon

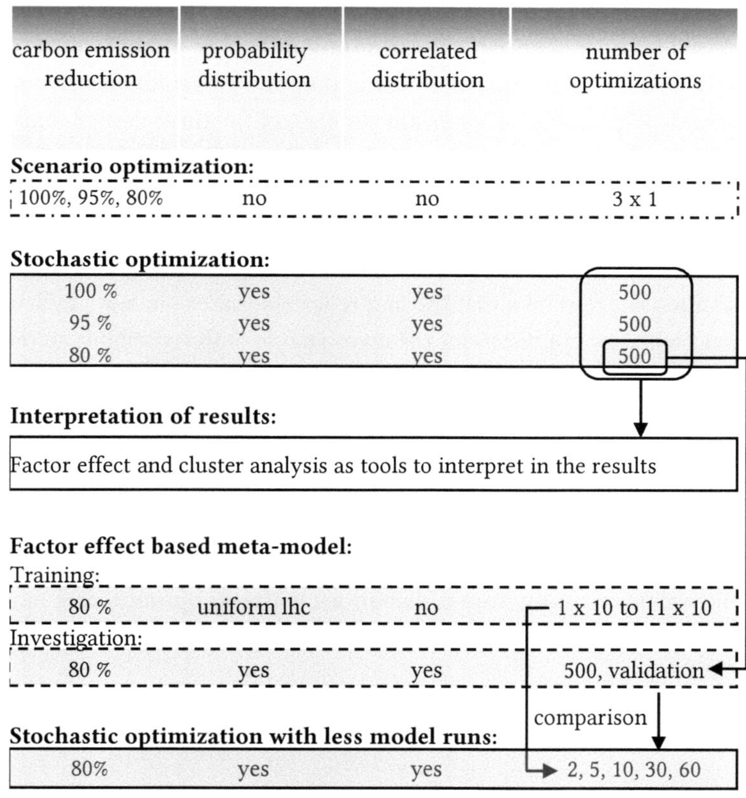

Figure 5.7: Optimizations and investigations carried out based on the ESO model described in this chapter.

emission reductions of 100 %, 95 % and 80 % compared to 1990 emissions from the energy sector without diffuse fuel emissions according to [3]. The 1990 benchmark is 968 Mt CO_2 emissions [3], the reduction thresholds therefore 0, 49.3 and 197 Mt. As preliminary optimizations on a simplified energy system model [95] have shown that 500 optimizations lead to a satisfactory representation of the output parameter distributions the probability distribution is discretized at 500 points. This means for each of the three carbon emission cases 500 optimizations are performed in order to simulate the continuous input parameter distributions. The computational effort to do so is relatively high, especially considering that more complex models, e.g. ones that take into account grids or neighboring countries, could be implemented. As preliminary optimizations have shown that the non renewable cases are more difficult to represent with few optimizations the investigation of the possibility to reduce computational demand and use a meta model is carried out with the 80 % carbon emission reduction boundary condition. The meta model is validated by calculating the mean squared error and the coefficient of determination. Based on these findings the box-plots of the predictions from the meta model with an desirable trade off between quality and computational effort are compared to those of the exact solution from the 500 benchmark optimizations, as well as to box-plots resulting from distributions with less optimizaions, namely two, five, ten, 30 and 60.

Chapter 6

Results

In this chapter the results of the different investigated variants are reported. One focus is on the general results of the optimization, i.e. which kind of energy system might be optimal in the future under the assumptions taken herein. The second focus is on the amount of information conveyed by each of the different approaches. Also the trade-off between computational expense and gathered information is investigated for the simplified stochastic approach in comparison to the extensive stochastic analysis. The parameters investigated are the total system cost, respectively the cost of energy, of a system concerning the overestimation caused by ignoring the uncertainty, as well as the optimal installed capacities of the different technologies. The metrics to account for the difference in information and to characterize the resulting distributions are the interquartile range, the mean, the median and the maximum and minimum values of each output parameter. Since the number of output parameters is too high to discuss all of them in detail, the analysis is limited to representative or, to the mind of the author, insightful results. The absolute cost of the energy system, as well as the respective cost of energy (CoE), resulting from a division of the total cost by the total amount of energy supplied, i.e. the sum of all hourly demand time series, is only used for the comparison between different system configurations, no claim as to its comparability to real world costs is made. This is because many aspects of the real energy system cost are too complex to be completely incorporated in the model system herein to draw conclusions from the resulting cost figures as e.g. power lines and the maintenance cost of district heating grids are completely omitted and represent an important cost component of the energy system. Also the actual current cost of the energy system is not known for comparison.

6.1 Results of the Scenario Approach

The underlying assumption herein is that a scenario with fixed values is not perceived as free of uncertainty but rather as a reasonable approximation by assuming the expected value, i.e. the mean of each uncertain parameters distribution as its value. Therefore, first the results from the scenarios, the scenario results (SR), that take the mean of each distribution as the parameters value, are discussed. Since uncertainty is not included explicitly in the optimization, results are single figures for each output parameter of the three different CO_2 emission limit scenarios. One first result is that the 80 % and 95 % emission reduction scenarios lead to identical results with an emission of 10.47 million tons of CO_2, making an emission reduction of 98.94 % compared to 1990 more cost efficient than using more of the CO_2 budget. One part of the explanation is the comparatively high carbon emission cost assumed herein. Together with the assumed technology cost reductions the optimal system is influenced in such a way, that higher renewable production together with carbon neutral storage is more cost effective than the use of more fossil methane with associated carbon emission costs. The possibility to use fossil methane in the energy system leads to a 0.3 % reduction in cost, reducing the CoE from 48.42 € MWh^{-1} in the completely renewable scenario to 48.27 € MWh^{-1}. Since the two not completely renewable scenarios have equal results, only two different results regarding the dimensioning of the technologies remain. In order to structure the analysis of the results the different technologies are discussed according to their belonging to the following groups:

- Renewable generation and fossil consumption

- Electricity storage

- Power to gas

- Gas storage

- Gas to power

- Power to heat

- Gas to heat

- Heat storage

As most of the resulting dimensioning of the technologies is equal or very similar first the 100 % renewable case is presented and then the differences to the non renewable case are highlighted. The installed capacities are rounded appropriately for readability.

6.1.1 Renewable Generation and Fossil Consumption

Regarding generation from renewable sources, both scenarios are comparable, as evident from the results listed in table 6.1. The optimal installed capacity of on- and offshore wind power plants lies at their respective imposed upper limits of 200 and 85 $GW_{peakpeak}$. In the case of photovoltaic installation, for which no limit is implemented in the model, the optimizer gives 2.19 TW_{peak} as the optimal peak power in the case of a completely renewable energy system (RES) and 2.08 TW_{peak} for the case of a not completely renewable energy system (NRES). This difference of about 5 % is, with 108 GW_{peak}, more than double the installed peak power of photovoltaic in Germany in the end of 2019 [161]. Both of the optimally installed capacities could be realized according to the potential areas for photovoltaic installation in table 5.7 on page 79. The fossil methane used in the not completely renewable case is 53.1 TWh which corresponds to 2.7 % of the total end energy demand in the system.

Table 6.1: Renewable generation technologies in the scenarios.

| | emission reduction scenario | | |
technology	100 %	95 % and 80 %	unit
onshore wind	200	200	GW
offshore wind	85	85	GW
PV	2186	2078	GW

Table 6.2: Power to gas technologies in the scenarios.

| | emission reduction of | | |
technology	100 %	95 % and 80 %	unit
PEM EC central	7.7	8.4	GW
AEL	1261.2	1190.2	GW
methanation	3.6	0	GW

6.1.2 Electricity Storage

The optimization yields that the use of the electricity storage technologies battery and redox flow battery are not cost optimal in the scenarios, thus these technologies are not used. Electricity storage is therefore realized via implementation of power to gas technologies and the respective electricity generation from gas to power technologies.

6.1.3 Power to gas technologies

From the in section 5.2.4 listed power to gas technologies the optimizer chooses to use all, except on-site PEM electrolysis. The resulting installed capacities used for hydrogen generation and methanation are displayed in table 6.2.

The main capacity for hydrogen production is alkaline electrolysis with a total power of 1261 GW_{el} in RES and 1190 GW_{el} in the NRES. The second most

Table 6.3: Gas storage in the scenarios.

| | emission reduction of | | |
technology	100 %	95 % and 80 %	unit
CH4 cavern	15.5	0	TWh
H2 cavern	466.5	446.5	TWh

important option is centrally applied PEM electolysis with 7.7 GW installed power for RES, in case of a NRES this is increased by 9 % to 8.4 GW_{el}. The produced hydrogen's conversion to methane only takes place in the RES, installing 3.6 $GW_{CH_4, lhv}$ of methanation capacity.

6.1.4 Gas Storage

In order to allow for the use of the produced gases at other times than their production, storage facilities must be included. The optimized storage for both cases is depicted in table 6.3, the values refer to the lower heating value of the gases.

The result shows that mainly hydrogen is stored, with a capacity of 466.5 TWh in the RES and 446.5 TWh in the NRES. A methane storage of about 15.5 TWh is only installed in the renewable energy system as in the NRES fossil methane has been implemented as commodity that can be supplied when it is needed. As there is not local hydrogen generation hydrogen storage from local generation in PEM electrolysis is not installed in the investigated scenarios.

6.1.5 Gas to Power Technologies

The conversion of the energy stored in hydrogen and methane to electricity is accomplished with PEM FCs, that are installed either central or on site, and combined cycle gas turbines for hydrogen usage. Only in the NRES scenario methane is used for electricity production. Methane SCGT and gas engines as

Table 6.4: Electricity generation from gaseous energy carriers in the scenarios.

| | emission reduction of | | |
technology	100 %	95% and 80 %	unit
CCGT CH4	0	11	GW
CCGT H2	162.5	150.9	GW
PEM FC central	7.7	8.4	GW

well as hydrogen gas engines and hydrogen SCGT are not part of the optimized scenarios. 11 GW of CCGT that use methane are installed in the NRES. The capacities of the three technologies that are applied for electricity generation from gas are depicted in table 6.4.

PEM FC on site generation with heat usage is not installed. The optimal amount of hydrogen CCGT lies at about 162.5 GW for the RES and 150.9 GW in the NRES. 7.7 GW of PEM FC for central application is installed in the RES case, compared to 8.4 GW in the NRES.

6.1.6 Power to Heat Technologies

Since the ESO model treats a renewable or almost completely renewable energy system the use of electricity for heat generation can be expected to be present. The results are listed in table 6.5.

The important role of electrical heating is highlighted by the usage of seven out of the eight available options for heat generation from electricity in the model. The only option not used are heat pumps (HP) for "old buildings" built before 1974. The results suggest that because of the lower performance of heat pumps in old buildings due to the higher temperature levels an application of heat pumps for these cases is not viable. A capacity of 355 and 348 GW heating rods in the RES and NRES respectively, are installed. Also in new houses the use of heating rods plays an important role with about 57 GW installed in both cases. Heat pumps in new houses with 11.8 GW_{el} for both cases, that can supply between 11.8 and 69 GW_{heat} depending on the

Table 6.5: Power to heat technologies in the scenarios. Old and new building refer to decentralized, domestic heat production at 60°C and 40°C respectively. For centralized demand the resistance heater supplies heat at <100°C while the electrode boiler produces heat at 100-500°C. Direct electric heating is centralized as well and produces heat at over 500°C.

technology	emission reduction of		unit
	100 %	95 % and 80 %	
heating rod, old building	355.2	348.3	GW
heating rod, new building	57.3	56.7	GW
HP, new building	11.8	11.8	GW
central HP	21	21.4	GW
resistance heater	14.4	15.2	GW
electrode boiler	21.6	19.3	GW
direct electric heating	100.3	101.3	GW

outside temperature, and heat pumps for the use in district heating systems with 21 GW_{heat} for the RES and 21.4 GW_{heat} for the NRES case are the more efficient heat generation options. The low investment cost heat generation in district heating networks is provided by 14.4 GW resistance heater in the RES, respectively 15.2 GW in the NRES. Medium and high temperature heat is also at least partly supplied via electricity with 21.6, respectively 19.3 GW for RES and NRES of electrode boilers and about 100 GW direct electric heating for high temperature processes.

6.1.7 Gas Heating Technologies

An other option for providing heat for industrial or residential heating purposes is the burning of hydrogen or methane. The different optimized capacities are depicted in table 6.6. Of the various options available for the different temperature levels and centralized and decentralized options only the natural gas boiler for medium temperature, i.e. 100-500°C, applications,

Table 6.6: Gas heating technologies in the scenarios.

| | emission reduction of | | |
technology	100 %	95 % and 80 %	unit
boiler mid CH_4	8.7	8.7	GW
condensing boiler, old	0	9.1	GW
direct gas heating H_2	15.7	10.3	GW

direct hydrogen heating for high temperature processes, i.e. >500°C, as well as the condensing boiler for "old buildings", i.e. decentralized demand at 60°C, are used in the optimized scenarios. In the NRES 9.1 GW of low temperature heating and 8.7 GW of medium temperature heating with methane are provided. In the RES no condensing boilers are installed and, as in the NRES, 8.7 GW boilers for medium temperature heat are installed. 15.7 GW direct gas heating are installed in the renewable energy system and 10.3 GW in the case of the NRES.

6.1.8 Heat Storage Technologies

Most of the heat in this model is supplied via electricity. Electricity storage is only present via power to gas in alkaline and PEM EC and subsequent power generation in hydrogen CCGT and via the reverse process in the PEM. This method of electricity storage leads to an electricity to electricity round trip efficiency between 14.7 and 38.6 %. Therefore a more efficient option is to store heat directly if the COP of the heating system is below 2.56, especially for short storage times. The different optimized heat storage options in the model are represented in table 6.7.

Storage demand is higher for high temperature liquid metal storage and district heating storage in the case of the NRES, whereas molten salt and the various different residential heat storage systems are slightly bigger for the completely renewable energy system. The greatest single storage demand comes form residential buildings before 1974, with over 1060 respectively

Table 6.7: Heat storage (hs) technologies in the scenarios.

technology	emission reduction of		
	100 %	95 % and 80 %	unit
central hs	371	376	GWh
liquid metal hs	524	532	GWh
molten salt hs	112	105	GWh
residential hs, new	16	16	GWh
residential hs HP, new	41	41	GWh
residential hs, old	1060	1006	GWh

1006 GWh of storage capacity. This is equivalent to about 39 kWh or 1.7 m^3 of hot water storage for each living unit in Germany built before 1974. For newer buildings the results are lower with a total of about 57 GWh of storage in new buildings. This amounts to an average of about 4 kWh or 345 liters of hot water storage in each new living unit, of which in total 72 % or 41 GWh are already covered with the assumed storage that comes included with the heat pump system. On the medium and high temperature side about 532 GWh of liquid metal storage are required in the NRES and 524 GWh in case of the RES. The need for heat storage at the medium temperature level, 100°C-500°C, is the lowest of the three temperature levels, at 105, respectively 112 GWh for the NRES and RES respectively.

6.1.9 Summary of the Scenario Results

Summarizing the scenario results there is little difference between the RES and the NRES as emission limits are not taken advantage of to the full extend. The energy system relies on on- and offshore wind power plants to provide energy. Both are implemented as much as allowed by the limits imposed. The remaining energy input is achieved with the installation of photovoltaic systems, in the order of 2 TW$_{peak}$. Batteries or flow batteries are not installed as the result of the optimization. Power to hydrogen allows the storage of energy, most of the hydrogen is produced in centralized facilities by about 1.2 TW of AEL, only about 0.6 % are installed as centralized PEM. Decentralized PEM is not utilized, consequently waste heat cannot be used to satisfy heat demand. Hydrogen is only stored in caverns, on site storage capacity is not implemented. Installed methanation capacity is small for the RES and zero for the NRES case. Methane, available after methanation or due to the use of fossil methane, is used for heat generation between 100 and 500 °C in both, the RES and the NRES, as well as for low temperature heat generation and electricity generation in CCGT in the NRES. Hydrogen is used for high temperature heat generation in both cases. Combined heat and power is produced in neither gas engines nor decentralized PEM fuel cells. Electricity generation from hydrogen is provided by about 8 GW of centralized reversible PEM fuel cells as well as between 150 and 163 GW of hydrogen CCGT. Heat generation is performed mainly by means of electric heating, with heating rods in combination with storage providing peaking and load shifting capacity especially in older buildings with higher temperature requirements, where heat pump performance is worse. Central heating and heating in new buildings is provided by heat pumps and resistance heaters. Medium temperature demand apart from what is being covered with hydrogen or methane and a part of the high temperature demand are also met by electric heating. In order to allow for load shifting of heating loads, about 1 TWh of hot water storage is needed in older buildings. In newer buildings only 57 GWh are implemented. District heating uses a storage capacity of about 370 GWh while high temperature relies on ca. 530 GWh of liquid metal storage. Of molton salt storage, which is used for the medium temperature range, only about 110 GWh are necessary, as the use of gaseous energy carriers already offers flexibility for the medium temperature heat demand.

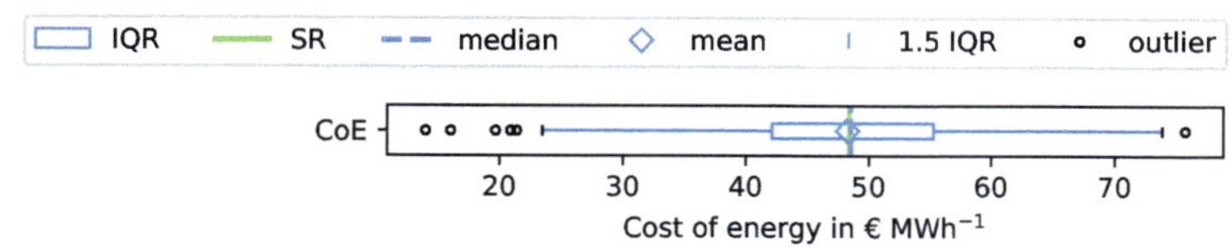

Figure 6.1: Stochastic optimization result for the cost of energy at 100 % emission reduction.

6.2 Stochastic Optimization

After the analysis of the mean scenarios, the results of the probabilistic investigation of the completely renewable case are presented. The results reported in detail in this section are limited to the renewable energy system, i.e. the system without fossil natural gas use. The distributions of the 95 % and 80 % emission reduction cases are recorded in appendices C.1 and C.2. Histograms, as used in chapter 4, are graphic, but impractical for the comparison of a higher number of distributions. Therefore box plots, marking the interquartile range, as well as mean, median and outliers outside of 1.5 times the interquartile range, are used for data visualization. For comparison, also the mean scenario result is marked in the box-plots.

The resulting distribution of the cost of energy is depicted in fig. 6.1. Regarding the comparison of system cost or cost of energy with the scenario result, an overestimation of the distributions mean by the mean scenario is expected according to Jensen´s inequality. The cost distribution´s mean lies at 48.21 € MWh^{-1}, the scenario result of 48.42 € MWh^{-1} represents an overestimation of expected system cost of 0.44 %. With a probability of 50 % the cost of energy as determined by the optimization lies between about 42 and 55 € MWh^{-1}. The upper 24.2 % of the distribution are between 55.2 and 70 € MWh^{-1} and the lower 23.6 % between 42 and 22 € MWh^{-1}. There are cases with a cumulative probability of 1% that the cost is lower than 22 € MWh^{-1} and four cases of the 500 optimizations where the cost lies above 70 € MWh^{-1}.

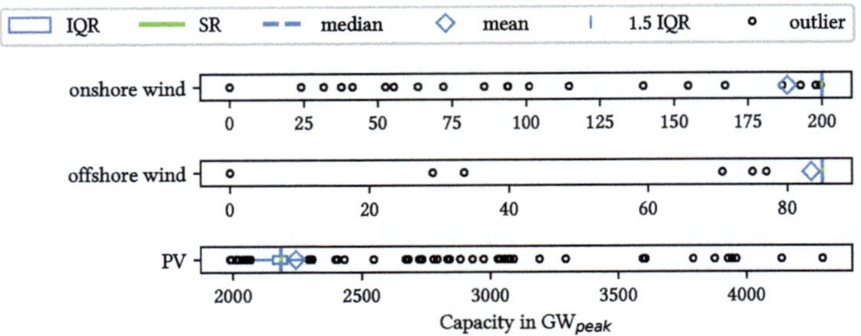

Figure 6.2: Stochastic optimization result for renewable electricity production at 100 % emission reduction.

6.2.1 Renewable Generation

Regarding renewable generation the presence of the fixed upper boundaries are determining for the distributions. The results are visualized in fig. 6.2. Interquartile range as well as median, and the 1.5 IQR collapse to a single point for land based and offshore wind. In the case of land based wind 8 % of the cases are below the limit of 200 GW_{peak}, with 4 % resulting in 0 GW_{peak}. Because of these outliers the distribution mean lies at 188 GW_{peak}. A similar pattern is discernible for offshore wind, here the cases below the limit of 85 $GW_{peakpeak}$ are just 2.6 % with a total of 1.6 % of the cases resulting in no installation of offshore wind power. Photovoltaic installation, for which no upper limit is imposed, shows a distribution with an IQR from 2.15 to 2.21 TW_{peak}. The median lies at 2.18 TW_{peak} and the mean at 2.24 GW_{peak}, the scenario result of 2.18 TW_{peak} is 2.7 % lower. The lowest amount of installed photovoltaic power plants in the optimized distribution is 1.99 TW_{peak}. Above the 1.5 IQR, that reaches up to 2.21 TW_{peak} there are outliers amounting to 24.2 % of the total number of optimizations. The distribution of the outliers is split, about 10 % of the outliers are between 3 TW_{peak} and 3.3 TW_{peak}, 82.6 % of ouliers, respectively 20 % of the total, lie between 2.21 and 3 TW_{peak} and 1.8 % of the total or 7.4 % of the outliers between 3.5 and the maximum 4.29 TW_{peak}.

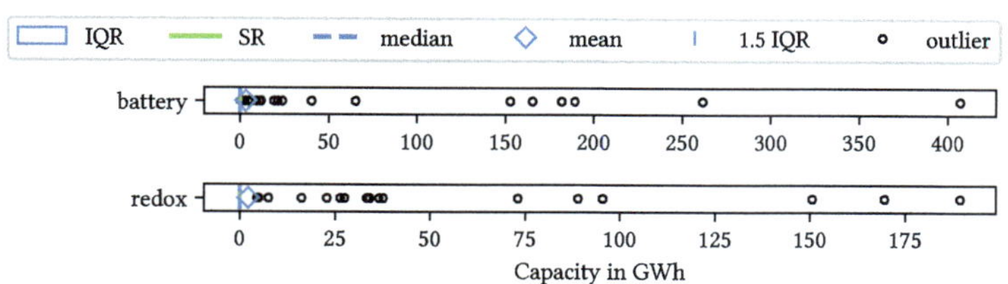

Figure 6.3: Stochastic optimization result for electricity storage at 100 % emission reduction.

6.2.2 Electricity Storage

The distribution of installed electricity storage is depicted in fig. 6.3. Differing from the mean scenarios, flow batteries are installed in some of the cases. The boxplot of conventional battery technology and redox flow battery collapses at 0 GWh installed capacity. The mean lies at 3.2 GWh, respectively 2.1 GWh in the case of redox flow battery capacity. For the battery this is driven by 19 of the 500 optimized combinations, representing a probability of 3.8 %. In case of the redox flow system 3.6% of the cases lie above zero installed capacity. The systems with either intalled battery systems or installed redox flow batteries do not have any installed capacity of the respective other electricity storage technology.

6.2.2.1 Power to Gas Technologies

Since no or little battery storage is installed, energy storage for the major part of the optimized configurations must at least partly take place via power to gas. The distributions resulting from stochastic optimization are presented in fig. 6.4. For PEM EC central the result is a distribution that is cut of at zero installed capacity to one side. The frequency of the results decreases from the resulting peak at zero installed capacity to contain 75 % of the cases between zero and 36.6 GW_{el}, 25 % of which between 12.4 and 36.6 GW_{el}. The mean is at 31.4 GW_{el}, its deviation from the median is driven by the fading out of the

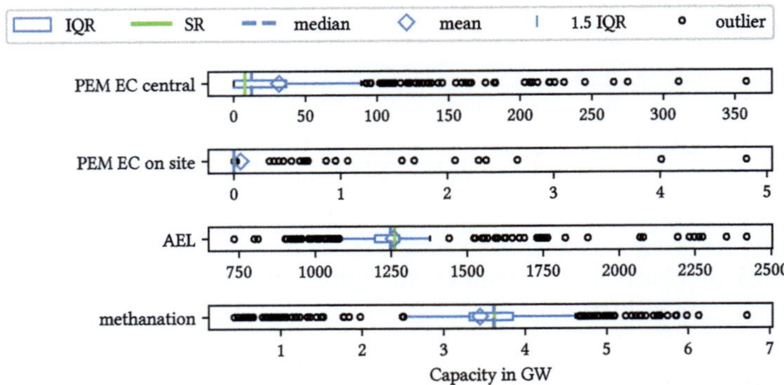

Figure 6.4: Stochastic optimization result for power to gas technologies at 100 % emission reduction.

distribution towards higher efficiencies with a maximum power of 357 GW$_{el}$. On-site generation with PEM electrolysis is only installed in 4.6 % of the cases. The mean of the cases when installed power is not zero lies at 1.3 GW$_{el}$, if all cases are considered the mean lies at 59 MW$_{el}$. The maximum amount of PEM EC for on-site generation is 4.8 GW$_{el}$. The distribution of optimized AEL installed capacity has an IQR of 78.8 GW from 1.19 to 1.28 TW$_{el}$. Mean, median and scenario result are close to one another, with 1.26 TW$_{el}$ for mean and scenario result and 1.25 TW$_{el}$ as the median. The AEL distribution is left leaning and has only 6.8 % of the cases above1.3 TW$_{el}$. While most of the outliers are lower than 1.8 TW$_{el}$, the maximum installed capacity of AEL is 2.4 TW$_{el}$.

Regarding methanation a closer analysis reveals a distribution with two peaks. One peak of the distribution is inside the IQR, from 3.3 to 3.9 GW$_{el}$, fading out towards higher capacities with a maximum of 6.7 GW$_{el}$. The lower peak lies between 2 GW and 0.4 GW$_{el}$. This split distribution leads to the high number of outliers visible in fig. 6.4.

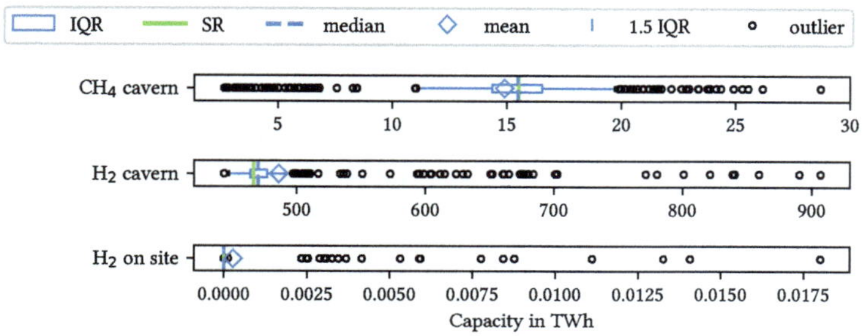

Figure 6.5: Stochastic optimization result for gas storage facilities at 100 % emission reduction.

6.2.3 Gas Storage

Dimensioning of gas storage in the completely renewable stochastic optimization is displayed in fig. 6.5. Similar to methanation discussed in the previous paragraph, cavern capacity for methane has a split distribution with two peaks, in the boxplot visible by the high number of outliers. The central peak between 12 and 17 TWh coincides mostly with the IQR from 12.7 to 15.8 TWh. The upper peak spans from 14 to 28 TWh and the lower from 2.6 to 7.5 TWh. The mean of the distribution is at 14.9 TWh, the median at 15.5 TWh. Hydrogen caverns, in all cases the greatest storage facility, has a left leaning distribution with a mean of 485 TWh and a median of 470 TWh. The IQR spans from 464 to 477 TWh, the minimum capacity of hydrogen storage required in the optimizations of the RES is 443 TWh. The most extreme case requires 907 TWh of installed hydrogen cavern storage capacity. The storage of hydrogen on-site happens on a smaller scale and in 94.8 % of the cases none is implemented. The mean of on side hydrogen storage is about 260 MWh, the highest storage capacity determined for on-site hydrogen storage is 18 GWh.

6.2.4 Gas to Power Technologies

From the available technologies for power generation, none of the CH_4 technologies, including CCGT and gas engines for central and on-site generation, are used in any of the optimized cases in the RES. From the hydrogen technologies neither of the gas engine technologies is implemented. The boxplots of installed capacities of technologies implemented in at least one of the distribution´s cases are depicted in fig. 6.6. In the case of SCGT using hydrogen, median and scenario result are at zero installed capacity and the mean of the distribution is at 3.5 GW installed capacity. The IQR ranges from zero to 3.8 GW, 25 % of the optimizations are distributed between this value and the maximum installed capacity of 30.7 GW. The expected value of hydrogen combined cycle gas turbine installation is 165 GW, the median is 164 GW. The distribution has one main peak around the mean and median values, with the IQR ranging from 163 to 165 GW. The outliers between 165 and 180 GW account for 20.6 %, those between 200 and the maximum 216 GW for 1.4 % of the optimizations. The, regarding the distribution mean, second most implemented power from gas generating technology is PEM FC in central application with a mean of 31.4 GW and a median of 12.4 GW. The distribution is the same of that of PEM EC central, as it is the same technology. This is also the case for PEM FC for on-site generation.

6.2.5 Power to Heat Technologies

Similar to the scenario results heat pumps for buildings built before 1974 are the only power to heat technology not applied. The box-plots of the other technologies installed capacities are depicted in fig. 6.7. Heating rods in old houses have a distribution between 253 and 418 GW, with an IQR between 333 and 366 GW. The mean is 353 and the median 354 GW. In newer houses the distribution is left leaning, with an IQR between 56 and 59 GW, the mean at 57.9 and the median at 58.3 GW. There is a peak between the minimum installed capacity of 54.2 GW and 55 GW, representing 17.2 % of the optimizations. The

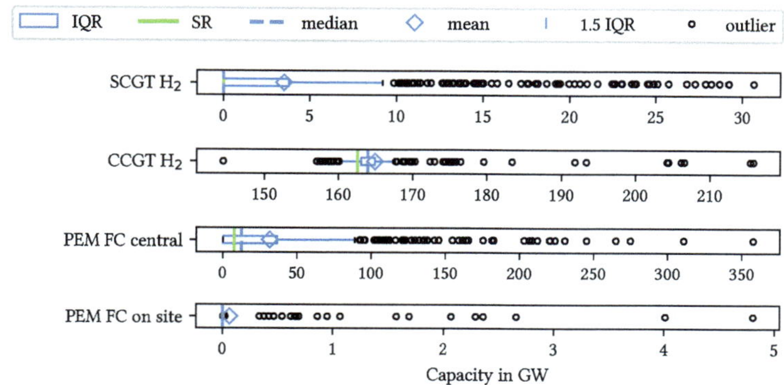

Figure 6.6: Stochastic optimization result for electricity generation from gaseous energy carriers at 100 % emission reduction.

maximum amount of heating rods in new houses in the distribution is 856 GW. Heat pumps in new houses are installed at a mean capacity of 11.6 GW, with an IQR from 11 to 12.4 GW. The median is 11.8 GW, some extreme cases allow only for a minimum installation between 0.5 GW and 5 GW. Central heat pumps in district heating networks are applied with a mean of 21 GW, the distribution having an IQR of 0.65 GW from 20.64 to 21.29 GW. The median is 21 GW, the system with the minimum installed capacity of central heat pumps deploys 13.5 GW. Resistance heaters in district heating networks are not installed in 2.6 % of the cases. Apart from the resulting peak at zero installed capacity the distribution is relatively symmetric with an IQR between 13.5 and 16.7 GW. The mean is 14.4 GW and the median 14.7 GW. The case where maximum resistance heater capacity is required has 40.6 GW installed. The electrode boiler for heat generation between 100 and 500 °C is installed in 50 % of the cases between 21.5 GW and 22.6 GW with the median at the boundary of the IQR at 21.5 GW. The mean, 22 GW, is slightly higher. The highest electrode boiler capacity used in one of the optimized systems is 25.3 GW, the lowest 16.1 GW. There is a peak representing 35 % of the cases, which have an

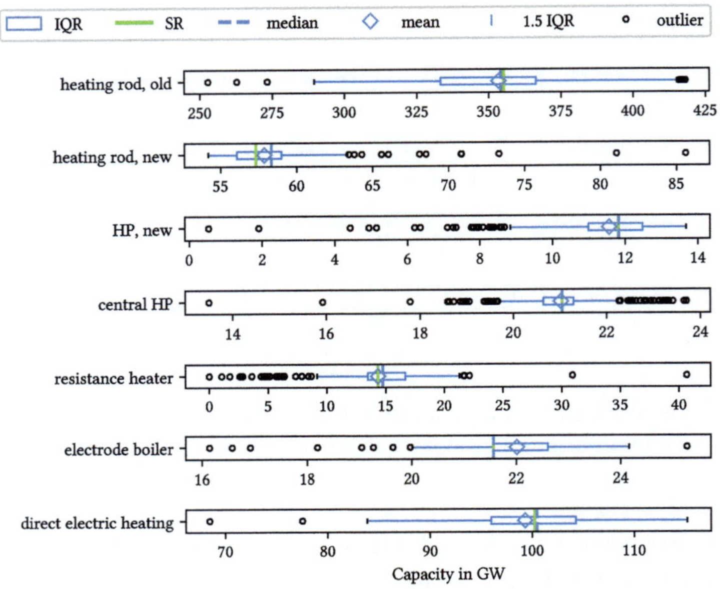

Figure 6.7: Stochastic optimization result for electric heating technologies at 100 % emission reduction.

electrode boiler capacity between 21.5 and 21.6 GW. Direct electric heating for high temperature process heat is installed with an IQR from 96 to 104.3 GW, a mean of 99.3 and a median of 100.5 GW. There are two outliers, systems in which 68.5 and 77.5 GW of direct electric heating are used.

6.2.6 Gas Heating Technologies

In case of the gas heating technologies five of the six available technologies are used in at least one of the optimizations, differing from the scenario result which employs only two of the five. The results of the stochastic optimization are pictured in fig. 6.8. The technology that is used neither in the scenario nor in the stochastic analysis is direct gas heating with methane. The least used technology in the stochastic optimization apart from this, is the methane boiler for district heating. In 98.4 % of the optimizations comprising the stochastic result, this technology is not implemented. The cases where it is implemented are discernible in fig. 6.8, with a maximum installed capacity of 2.2 GW. One of the technologies providing heat in scenario and stochastic results is the methane burning boiler for the temperature range between 100 and 500 °C. In the stochastic optimization its distributions IQR ranges from 8.1 to 8.7 GW installed capacity. The mean is 8.2 GW, close to the lower end of the IQR, the median case is 8.7 GW, coinciding with the upper end of the IQR. A closer analysis of the distribution shows three sharp peaks at 7.5 GW, 8.1 GW and 8.7 GW, with each of these incorporating in an interval of 0.1 GW 9.2 %, 11.8 % and 59.6 % of the distribution respectively. 1.2 % of the cases yield results below 1 GW with a minimum of no installed capacity. In 79.6 % of the cases condensing boilers in houses build before 1974 are not a part of the technologies used for heat generation. Therefore the box-plot collapses at zero installed capacity, falling together with the scenario result. The mean lies at 2.4 GW, the maximum installed capacity is 32.4 GW. Hydrogen powered gas boilers for the medium temperature range are not installed in 93.2 % of the cases. The mean lies at 117 MW, the maximum installed capacity is 8.7 GW. In case of direct gas heating with hydrogen, 28.6 % of the optimizations result in a system where this technology is not used. The rest of the results are distributed between zero installed capacity and the maximum value 24 GW. 46.8 % of optimization results are distributed between 10 GW and 20 GW, 23.8 % between 0 GW and 10 GW . The mean of the distribution is at 8.5 GW, the median is 8.8 GW.

6.2.7 Heat Storage Technologies

The importance of heat storage is highlighted by the fact that for each of the five heating demand time series at least one storage technology is applied.

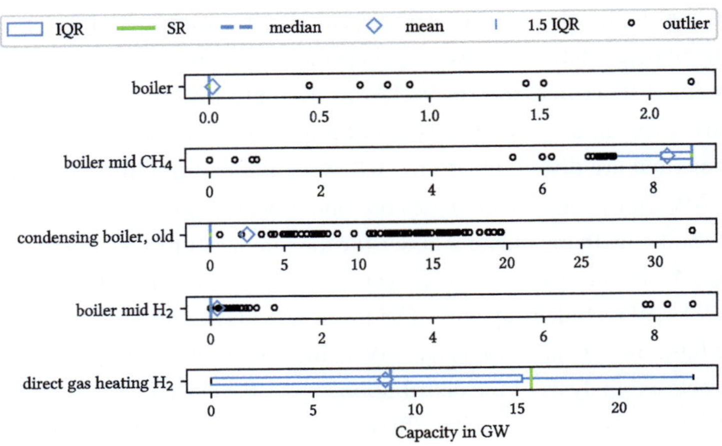

Figure 6.8: Stochastic optimization result for heating with gas at 100 % emission reduction.

The resulting distributions are depicted in fig. 6.9. For district heating the implemented storage is central heat storage. The distributions IQR is between 342 and 406 GWh. The mean is at 382 GWh, the median at 375 GWh. Maximum capacity of central heat storage is 644 GWh. Liquid metal storage, allowing the decoupling of high temperature heat production from demand is installed with a mean capacity of 517 GWh. The IQR ranges from 500 to 536 GWh. The minimally installed liquid metal storage is 324 GWh. The expected capacity of necessary molten salt storage for the medium temperature range up to 500 °C is 116 GWh, the IQR spans from 112 to 122 GWh, with the median at the lower end at 112 GWh. The distribution contains four peaks at 113, 119, 122 and 128.5 GWh, which contain within a range of 1 GW 49.6 %, 8 %, 12.6 % and 5.8 % respectively. Residential, decentralized heat storage in new houses comes included with the heat pump and can be added separately. For the separate storage the mean lies at 16.4 GWh with an IQR from 10 GWh to 22.9 GWh. In 2 % of the cases no additional heat storage to that included with the heat pump

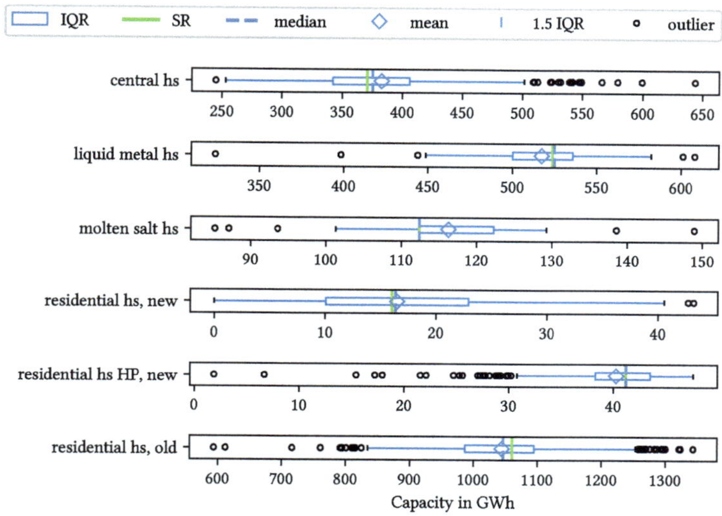

Figure 6.9: Stochastic optimization result for heat storage at 100 % emission reduction.

is needed in new houses. As the distribution of heat storage included with the heat pump is proportional to the heat pumps distribution it is not explained in detail at this point. Houses built before 1974 have an expected storage demand of 1044 GWh resulting from the stochastic optimization, with the optimal installed capacity in 50 % of the cases lying between 987 and 1094 GWh. The minimum optimal heat storage installation in older buildings is 594 GWh, the maximum 1344 GWh.

6.2.8 Summary of the Stochastic Optimization Results in the Renewable Case

The probabilistic investigation shows a broad field of different possible energy systems. The IQR, i.e. 50 % of the frequency, is condensed to one point of the distribution in the 100 % renewable system e.g. where fixed upper boundaries such as in the cases of on- and offshore wind power exist. Another case where the whole IQR is condensed to one point can be at zero installed capacity, e.g. in the case of batteries, redox flow batteries, PEM EC on-site, hydrogen storage on-site and condensing boilers in old houses, as well as boilers in central heating and hydrogen boilers for the medium temperature range. For all of the latter there are at least some cases in which the technology is implemented in the GW or GWh range, while for the former there are also cases in which no off- or onshore wind is implemented. Looking at the system perspective the 100 % renewable energy system can be said to have a probable cost in the range between 42 € MWh^{-1} and 55 € MWh^{-1} under the taken assumptions and with the limitations regarding the interpretation of system cost mentioned. Renewable electricity generation will probably be cost optimal with 200 GW$_{peak}$ onshore wind and 85 GW$_{peak}$ of installed offshore wind power, while the expected value for these is lower, because there is the possibility for cost optimal systems with little or none of either technology. Renewable electricity generation from photovoltaic power plants is important in every system configuration resulting from the stochastic optimization. It is between 1.99 and 4.29 TW$_{peak}$, and probably between 2.15 and 2.21 TW$_{peak}$. The expected optimally installed capacity, i.e. the distribution mean is 2.24 TW$_{peak}$. Providing electricity storage is probably cost optimal via power to gas processes, here especially alkaline electrolysis at about 1.2 to 1.3 TW installed power and a minimum of 750 GW plays an important role. Methanation as well as methane storage in caverns is necessary in every cost optimal system investigated, yet the importance varies between less than 1 GW and more than six, as well as between four and over 25 TWh of storage capacity. Hydrogen storage in caverns is the most important energy storage system with a probable contribution of 464 to 477 TWh, but an implementation of as much as 907 TWh can be cost optimal. The produced gases can be used for electricity production in hydrogen powered SCGT or CCGT. In these cases the main contribution comes from CCGT with between 160 and 170 GW installed capacity while SCGT only contributes in some cases. With a probability of 75 % less than 10 GW of SCGT is optimal. The implementation of reversible PEM FC varies between no installed capacity in about 27 % of the cases and over 350 GW, with a probability of 25 % that the optimal capacity is between 12.4 and 36.6 GW. Heat is supplied to old houses mainly with heating rods in the houses, of which probably about 330 to 370 GW are optimal. Heat pumps are never used in old houses, the use of condensing boilers in these is in over 75 % of the cases not cost optimal, yet there are a number of cases with up to 20 GW implemented. Heat pumps play an important role in new houses and for district heating with 10 to 14 GW installed in new houses and about 20 to 22 GW in district heating networks. Here also 10 to 20 GW resistance heaters are implemented. Heat pump usage in new houses is complemented by about 55 to 60 GW heating rods. Medium temperature heat is provided by 20 to 24 GW electrode boilers as well as methane boilers, mainly between 7 and 9 GW, and in less than 25 % of the cases hydrogen boilers. High temperture heat is provided with direct electric heating between 90 and 110 GW as well as, in over 70 % of the cases, some direct heating with hydrogen. Direct heating with hydrogen is in about 50 % of the cases implemented between 10 and 20 GW. Heat storage is present for all temperature levels and in all cases except some where new houses do not need heat storage, which lies between 10 and 23 GWh for most cases, even if there are systems in which more than 40 GWh are optimal. Central heating networks probably make use of 340 to 410 GWh of stroage capacity. Medium temperature storage, namely molten salt heat storage, is probably implemented between 110 and 125 GWh. High temperature heat is stored in 450 to 580 GWh of liquid metal storage, with 500 to 540 GWh representing the most probable cases. The greatest heat storage capacity is hot water storage in houses build before 1974. Between 0.8 and 1.3 TWh are very probable with a probability of 97.6 %. With over 60 % probability the cost optimal amount is between 0.95 and 1.1 TWh.

6.3 Patterns within the Stochastic Optimization Results

Regarding the optimization results there are different types of questions one wants to answer with an analysis, depending on who performs the analysis for which reason. Some of these aspects are listed below.

1. What is the probable optimally installed capacity of each technology?

2. How much will it probably cost to supply energy under these assumptions?

3. Which technologies are most likely to play an important role in the future energy system?

4. Which technologies implementation is most or least dependent on other technologies cost developments respectively the present uncertainty?

5. The cost of which technology is especially influential on the future energy system cost and its structure?

6. What type of energy system layout is probably optimal for the future, i.e. what is the probable cost optimal system structure?

Some of the questions above, as e.g. aspect one and three can already be answered based on the box-plots presented in section 6.2. Others, such as aspect two are outside of the scope of a simplified ESO, the calculated cost of energy is only valid for comparison between different cases of the same model. Question four, regarding the influence of uncertainty on the specific technologies, can for example be relevant to policy makers that want to know which technology is always important in the energy system, or companies investing in a technology trying to estimate the risk of a certain technology becoming obsolete.

In order to answer some of the questions above two different approaches to uncover patterns in the optimization results are applied. These are firstly the factor effect analysis, aiming at identifying the influence each single factor, herein the cost of a technology, has on the results. This methodology is explained in more detail in the next section. The second approach is based on the use of a clustering algorithm to identify clusters of similar systems.

6.3.1 Factor Effect Analysis Applied to the Stochastic Optimization Results

The before mentioned questions regarding probability of obsolescence, as well as which technology implementation or output parameter depends on which other technology, can be answered by using an aspect of factor effect analysis that captures the underlying interactions [103]. The herein used factor effect analysis is based on [103]. By testing different system configurations, i.e. input parameter settings, the influence of each input parameter on the system behavior, in this case e.g. total system cost of the optimally installed capacity of a technology, can be calculated. The basic idea is that each factor, herein technology cost, has an influence on the output parameter. By adding the systems mean to these influences in terms of the slope of the linear fit as seen in fig. 6.10, multiplied with the factor value, an approximation of the output variable depending on the factor values, i.e. the technology costs, can be calculated as shown in eq. (6.1) from [103]. The meaning of the variables is the following. y is the so called quality aspect, e.g. system cost, that we want to approximate. c is a model constant, herein c_0 is the mean of the quality aspect in the training data, c_i is the linear slope of the change of the quality aspect with the respective input variable x_i. c_{ij} is the linear slope of the change of the quality aspect over $x_i^* x_j$. ε is the deviation of the model equation from the value of the actual quality aspect.

$$y = c_0 + \sum_{i=1}^{n_f} c_i x_i + \sum_{i=1}^{n_f-1} \sum_{j=i+1}^{n_f} c_{ij} x_i x_j + \varepsilon \qquad (6.1)$$

Interactions between two input variables (e.g. effect of higher PV cost with lower offshore wind power cost) are expressed by the third term on the right hand side in eq. (6.1). Also quadratic or cubic interactions can be used. Quadratic effects can be represented by linear slopes if the quadratic value is directly assigned to the first axis. The same holds for higher order polynomials, resulting in the possibility to represent the whole system as a sum of linear functions.

An exemplary result of factor effect analysis for the three investigated emission cases and the main factors, namely the uncertain technology costs, is presented in fig. 6.10. The dotted lines represent the best linear fit of the data visualized in the scatter plots. The horizontal axis displays the relative cost of the technology in the range between 2018 cost from section 5.2 as the upper boundary and zero cost as the lower boundary. Note that redox flow battery input parameters have been excluded, as redox flow batteries are only applied in few of the optimizations. The represented linear fits are the basis of regular factor effect analysis. The sum of the factor effects, i.e. the slope of the linear fits, each multiplied with the relative factor value, yields a linear model that is suitable to describe many systems [103]. The extrapolation of the linear fit visible in fig. 6.10 and the subsequent figures is only for visualization purposes. The extrapolated range should not be used to construct a model and make predictions outside of the range used for model parametrization. To take into account quadratic or higher order effects, a linear representation can be achieved by representing the squared factor value on the horizontal axis. Interactions between two factors can be accounted for following the same principle [103]. The strength of factor effect analysis is that it is a white box model, i.e. a model whose inner workings can be understood and interpreted by the user. The factor effects are the effect the factor depicted on the horizontal axis has on the investigated result variable. All other factors influence is averaged over its parameter range, thus revealing the influence of the depicted factor.

Figure 6.10 reveals challenges and advantages of using factor effect analysis. On the one hand the use of distributions and fits reveals the magnitude of influence a technology's cost has on the quality aspect, in this case the CO_2 emissions. The influence of PV cost and of the decision factor, the CO_2 limit, is discernible in the upper left graph, showing that with increasing CO_2 emission limit the effect of increasing PV cost is amplified. Regarding the upper left graph, it is clear that a linear, or even polynomial fit of high order cannot approximate a discontinuity such as the step observed herein. This is the limitation of the use of conventional fitting and necessitates the use of kriging or artificial neural networks [103]. Another important aspect regarding the interpretation of fig. 6.10 is, that the herein analyzed data has underlying probability distributions that are correlated according to the use of equal subsystems or component groups as given in table 5.20. Therefore some of the trends discernible can result from the correlation of costs, not from a parameter's influence.

In the example, fig. 6.10, the influence of uncertain technologies cost on the systems carbon emissions are discernible. The greatest influence is exerted by the photovoltaic costs. Below a threshold of about 40 % of 2018 cost, carbon emissions become zero independent of the allowed system emissions. Above this value there is a steep increase in carbon emissions between 50 and 60 % of 2018 PV cost, which with the 95 % emission reduction boundary condition levels out at the corresponding value, while in the 80 % case carbon emissions rise with a decreasing slope, staying below the 197 million ton annual carbon dioxide emission limit. An increase in cost of reversible PEM fuel cell, molten salt and metal storage as well as AEL and battery costs leads to an increase in carbon emissions as well. Even though the factor effect of e.g. molten storage or PEM cost are clearly discernible, the smaller assumed uncertainty for these cases reduces the actual influence of these costs on the emissions,

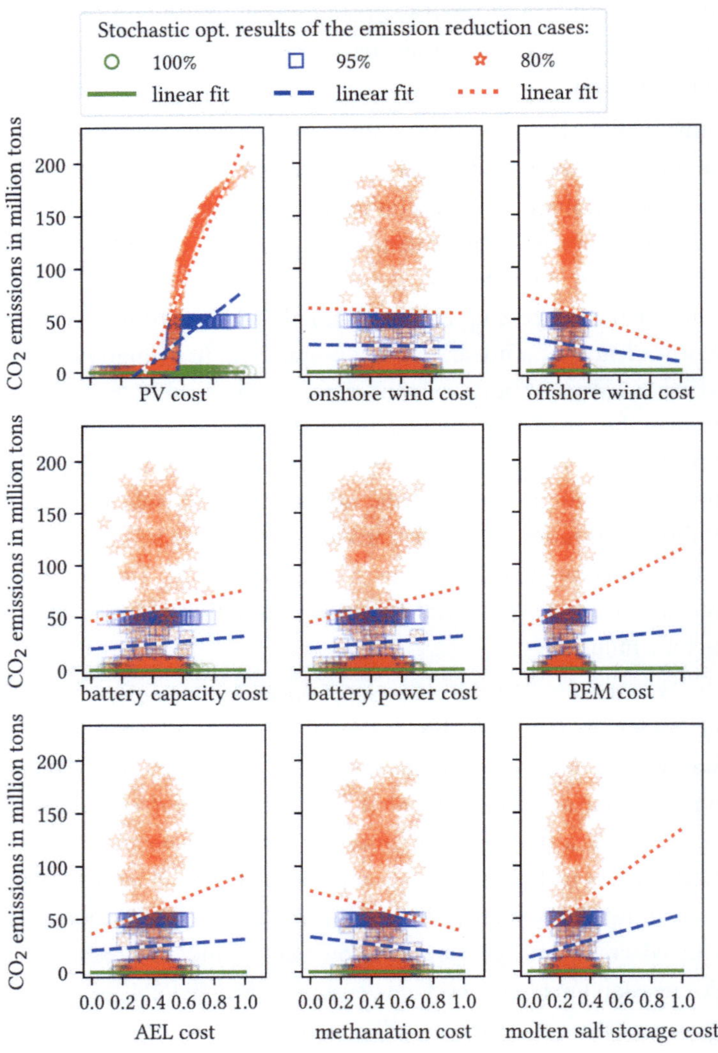

Figure 6.10: Effects of the uncertain technology costs except redox flow battery on carbon emissions and respective least square linear fits for the three emission limit cases. Extrapolation of the linear fits only for trend visualization. Horizontal axis range is relative technology cost between zero and 2018 cost from section 5.2.

as variations are smaller. Offshore wind power plants and methanation cost both behave inverse in relation to the carbon emissions, with decreasing mean carbon emissions at rising costs. A possible explanation of the mehanation cost's influence could be the double use of equipment when more renewable methane is produced, making e.g. CCGT that use methane more economically viable and, vice versa, reduce its viability if less synthetic methane is produced. In the case of offshore a possible explanation is revealed by a closer analysis of the results for the 80 % emission reduction case. All the cases where less than the maximum allowed amount of offshore wind power is installed are completely renewable, i.e. there are no carbon dioxide emissions. In these cases photovoltaic installations are between 2.5 and 4.5 TW_{peak}, representing the upper end of the PV installation range, which corresponds to the cases with below average costs for photovoltaic power plants. As decreasing photovoltaic costs have a strong effect on lowering system emissions, the counter intuitive result of lower carbon emissions at higher offshore wind power plant costs emerges. Here again it is important to note that these seemingly plausible explanations as to cause effect relations are not certain, as the complexity of the system makes a clear identification of cause effect relations impossible.

A technology example for the use of factor effect analysis is given in fig. 6.11. The application of reversible PEM FC for central application in the energy system is evaluated based on the influence of the uncertain technology costs, again redox flow battery power and capacity costs are excluded from the analysis for simplicity.

The factor effects for a technology can give a basis for answering if a technology's implementation is more dependent on its own cost development or that of other technologies that might have possible synergies or offer an alternative, if the before mentioned limitations as to cause effect relations are kept in mind. From fig. 6.11 it becomes clear that even though PEM fuel cell implementation in central applications depends heavily on PEM fuel cell cost, there are other notable influences, namely PV cost, molten storage cost and offshore wind power plant cost on PEM implementation. The influences of all

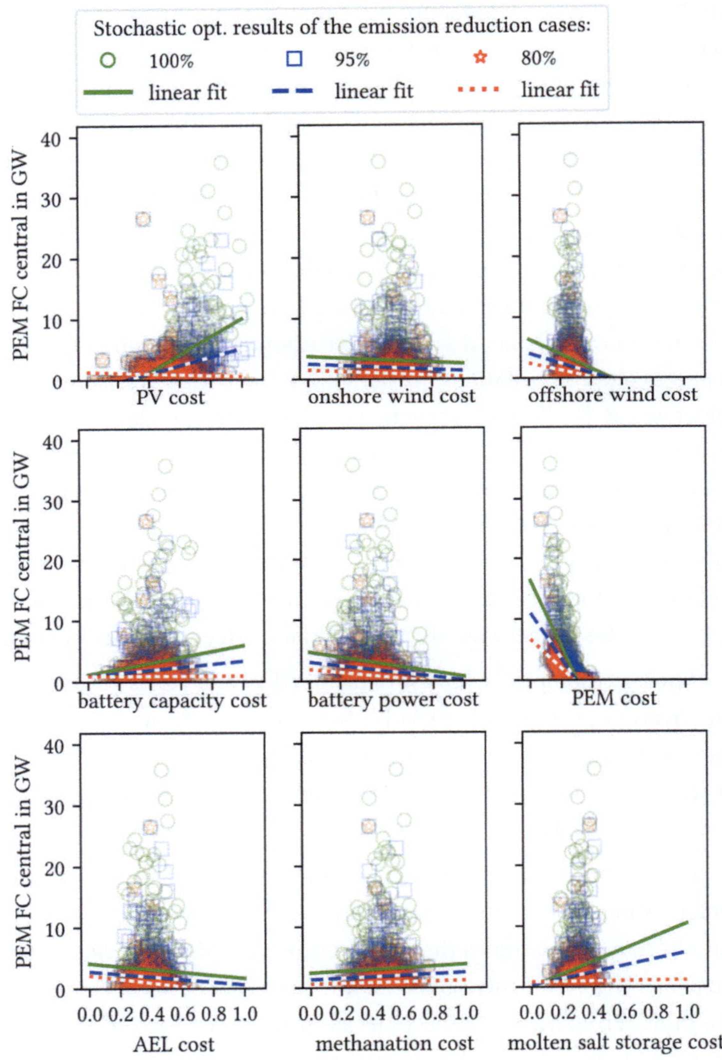

Figure 6.11: Effects of the uncertain technology costs except redox flow battery on the implementation of PEM fuel cells in centralized application and respective linear fits for the three emission limit cases.

the parameters are stronger the stricter the emission boundary conditions are. While PV cost does not influence the PEM FC central implementation if only 80 % emission reduction are required, in the completely renewable case more costly PV corresponds to higher PEM implementations presumably because fossil methane lowers the need for hydrogen production and re-electrification as a means of storing energy in the not completely renewable cases. The same is true for molten salt and metal storage, which indicates the energy storage role of PEM FC systems and a possible, at least partial substitution with molten storage when storage costs are low.

A similar analysis can be performed for all the implemented technologies. The use of factor effects to construct a meta model and predict optimization results is explored in section 6.4.2.

6.3.2 Clustering of the Stochastic Optimization Results

Regarding the last two questions from the beginning of this section which concern the structure of the energy system, it is not enough to consider box-plots or factor effects of each technology, as the information about how the system is structured is not contained within this information. Stochastic optimization results are more difficult to interpret, due to their distribution nature and the dependencies between all parameters which are captured. Another possibility for data analysis is the use of clustering algorithms. Clustering algorithms allow assigning different objects to collectives, which are based on the similarity of the members and the differences between members of different collectives [240]. Herein the clustering algorithm *hdbscan* described in [241] is used. It combines the advantages of hierarchical and density based clustering, enabling the extraction of clusters from different levels of a cluster tree and allowing for the possibility that these are characterized by different density thresholds. Another advantage is the possibility of using the algorithm with a single input parameter, i.e. the minimum number of objects for a cluster to be defined as a cluster, allowing for results that do not depend as heavily on user input as other clustering algorithms. [241]

The algorithm is implemented as a python package [242]. A basic problem of clustering is that clusters can become very general if too much data is put into one cluster, e.g. all the data in one cluster in the extreme case, such that there is no added value from clustering anymore. The other side of the coin is that clusters become more meaningful if more clusters are introduced, but there is the problem that also many clusters become meaningless if, in the extreme case, each item receives its own cluster. Therefore herein optimization of the minimum cluster size is performed by minimizing the amount of data with a low probability, herein smaller 5%, of belonging to one of the clusters as proposed by [243]. This yields the minimum cluster size for each data set. A second condition for this optimization is set in this work. The maximum number of clusters must not be greater 20 in order for the clusters to retain planning and interpretation value for the user. The resulting clusters for the 100 % renewable energy system are presented in figs. 6.12 to 6.14.

In the case herein we cluster the different energy systems resulting from the stochastic optimization, i.e. 500 different system configurations, by their similarity regarding technology implementation. The degree of similarity and where to draw the line between different clusters is determined by the algorithm. The clustering yields two clusters, *group 1* and *group 2*, as well as the *group 0* of elements that do not fit in either of these. Figures 6.12 to 6.14 are a representation of the output parameters. These are, together with the installed charge and discharge power of each storage technology which is not represented in the plots, the basis for the cluster definition. The two clusters and *group 0* are distinguishable by the optimal technology installations. This is an advantage of the chosen spider or radar plots. Electricity generation is easily discernible as almost solely PV based for *group 0*, PV, onshore and offshore wind power based in *group 1* and PV and mainly offshore wind power plants based in *group 2*. More differences are revealed between the clusters when looking at storage technologies. *Group 1* uses 40 to 50 % of the maximum

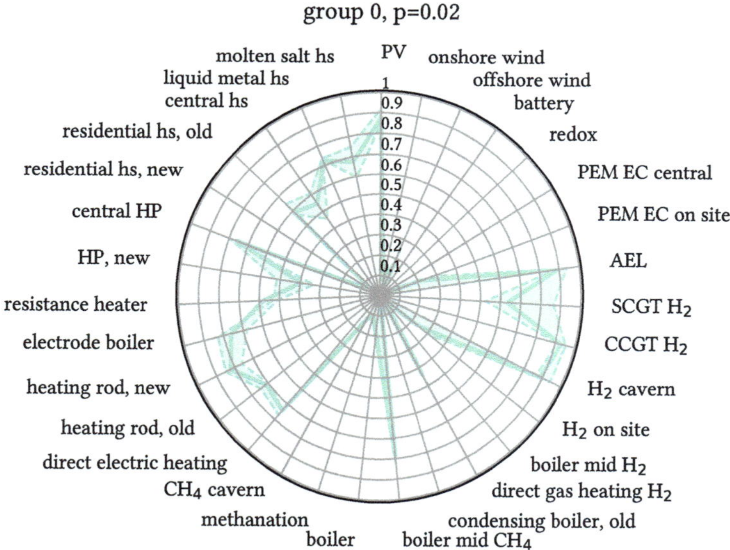

Figure 6.12: *Group 0* resulting from the clustering of the stochastic optimization in the 100 % emission reduction case. The radial axis displays the relative amount of optimized technology in the range between zero and maximal installed capacity of each technology in the given emission scenario, the solid line shows the mean values, dashed lines depicts min and max of the IQR, which is colored.

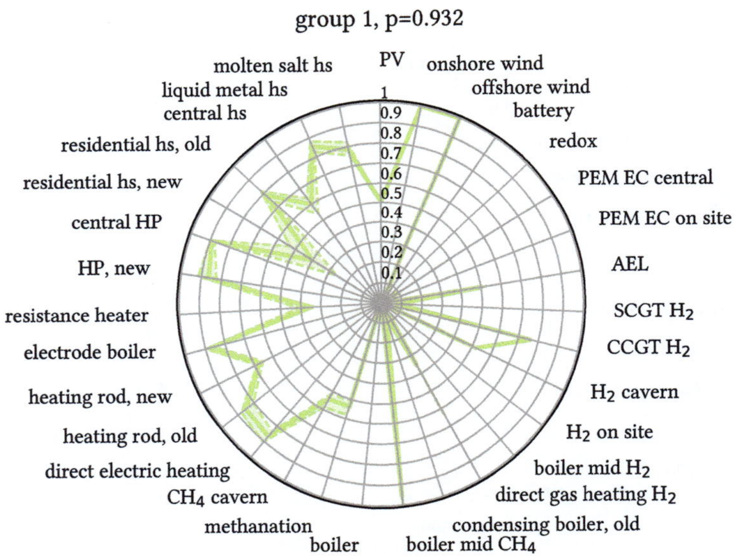

Figure 6.13: Cluster resulting from the clustering of the stochastic optimization in the 100 % emission reduction case. The radial axis displays the relative amount of optimized technology in the range between zero and maximal installed capacity of each technology in the given emission scenario, the solid line shows the mean values, dashed lines depicts min and max of the IQR, which is colored.

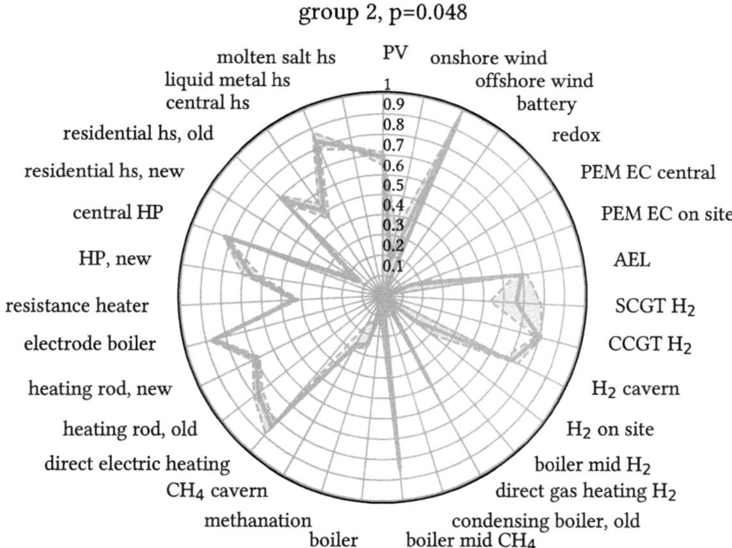

Figure 6.14: Cluster resulting from the clustering of the stochastic optimization in the 100 % emission reduction case. The radial axis displays the relative amount of optimized technology in the range between zero and maximal installed capacity of each technology in the given emission scenario, the solid line shows the mean values, dashed lines depicts min and max of the IQR, which is colored.

amount of methanation and cavern storage for methane used in all systems while *group 2* only applies 10-20 %. Also the use of heat pumps and heat storage in new building plays a more important role in *group 1* then in *group 2*. One question regarding these different system layouts is which layout becomes more probable under which conditions regarding the input parameters, i.e. the technologies' costs. In order to allow answering this, the mean of each technology cost for the three groups with their respective probabilities are presented in table 6.8. Additional to the input parameters also the mean cost of energy of each cluster is listed in the table. For comparison the mean of all 500 optimized cases is displayed as well. One observation that is possible from this data is e.g. that *group 2* system layouts result from cheaper PV, with a mean of 75 % cost reduction compared to only 45 % in the mean *of group 1* systems, combined with more expensive onshore wind power plants, with only 37 % mean cost reduction compared to 45 % in the case of *group 1*.

Table 6.8: Relative cost mean of input parameters and cost of energy in € MWh^{-1} by cluster in case of 100 % emission reduction.

	all cases	group 0	group 1	group 2
cluster probability		0.02	0.932	0.048
photovoltaic	0.53	0.14	0.55	0.25
onshore wind	0.55	0.53	0.55	0.63
offshore wind	0.27	0.26	0.27	0.29
battery, capacity	0.41	0.41	0.41	0.41
battery, power	0.41	0.42	0.41	0.39
PEM	0.24	0.23	0.24	0.24
alk. electrolysis	0.40	0.42	0.40	0.39
methanation	0.47	0.54	0.46	0.47
redox power	0.52	0.51	0.52	0.48
redox capacity	0.24	0.24	0.24	0.22
molten strorage	0.29	0.25	0.30	0.28
cost of energy	48.2	22.97	49.53	33.29

Cluster analysis can be performed for the 95 and 80 % emission reduction cases in the same manner. The resulting clusters and the table regarding the corresponding input parameters and cost of energy as well as carbon emissions for the 95 % emission reduction case are depicted in appendix C.3 in figs. C.19 to C.22 and table C.1. The resulting clustering with the respective probabilities for the 80 % emission reduction case are displayed in appendix C.4 on figs. C.24 to C.26. The clustering leads two clusters and again *group 0* with the cases that cannot be assigned to either of the clusters. The mean of the input parameters, cost of energy and carbon emissions for each cluster in the 80 % emission reduction case is displayed in table C.2.

6.3.3 Summary of Patterns within the Stochastic Optimization Results

The complexity of the data resulting from stochastic optimization, in this case the implementation of over 70 technologies in 500 cases with different input parameters, suggests the use of additional tools. Factor effect analysis as a white box model allows to understand the implication on parameter, e.g. photovoltaic cost, has on output parameters of the system, i.e. the system cost and the optimally installed capacity of each technology. In order to gain these insights the slopes of the linear dependency between the input parameter and the output parameters can be computed. The result gives a fast understanding and overview which technology cost change interacts positively o negatively with which other technology. The limitation is the representation of discontinuous behavior, which can not be discerned.

Clustering enables the researcher to identify different archetypes of energy system configurations based on the capacity of the installed technologies and assign probabilities to these. This highly reduces the complexity as not 500 different configurations have to be compared. The graphical representation with spider plots allows for a quick overview and insight which technologies always play an important role and how they play together to achieve a viable energy system.

6.4 Stochastic Optimization with lower Computational Demand

In order to be able to take advantage of the increased information gain from stochastic optimization also when dealing with models with high computation time, stochastic optimization with less optimization runs is needed. One possibility is a discretization of the probability distributions at fewer points. As preliminary investigating have shown that at small sample sizes the results are not necessarily representative of a distribution discretized at a high number of points [95], the use of a meta model is explored as an alternative approach. The investigations are performed for the 80 % emission reduction case, the results from the 500 optimizations presented above are the benchmark for the evaluation.

6.4.1 Stochastic Optimization with less Optimization Runs

The comparison is carried out for the box-plots resulting from stochastic optimization based on 60, 30, ten, five and two optimizations with the benchmark box-plot based on 500 optimizations. First exemplary box-plots of the distributions resulting are presented. After this the use of clustering in the considered cases is investigated.

Benchmarking of Box-Plot Information with less Optimizations

Exemplary box-plots for comparison are depicted for the cost of energy in fig. 6.15, hydrogen cavern storage in fig. 6.16 and PEM $FC_{central}$ in fig. 6.17.

The distribution for the cost of energy shows Jensen´s inequality and that it is better approximated with higher number of optimizations. When 30 or more optimizations are used the result is comparable to that of the benchmark. Mean and IQR are approximated relatively accurately with three or more optimizations, the position of the median shows that the shape of the distributions differs already between 60 and 500 optimizations. Presence and position of the outliers show expected behavior, less probable configurations represented by these become less common when fewer optimizations are used. Therefore outliers are underrepresented when a reduced number of optimizations is used. Summarizing the use of five optimizations gives an impression of the width of the IQR, while using 30 allows approximating the effect of Jensen´s inequality.

Regarding hydrogen storage capacity, width and position of the IQR are approximated well with 30 or more optimizations. This is also true for the mean, while the position of the median and subsequently the shape of the distribution needs 60 optimizations to be represented. The width of the whiskers, representing the outer parts of the distribution is greater for a greater number of optimizations, resulting from the non-representation of less probable, extreme cases when few optimizations are used. For PEM fuel cell in central application similar trends can be discerned, outliers and distribution width are underrepresented with fewer optimizations. In the case of PEM fuel cells already 30 optimizations exhibit a greater deviation from the benchmark result, with the IQR only covering about half of the benchmark's width.

In table 6.9 the results of clustering the optimization results in the three clusters obtained by clustering the benchmark optimization is depicted. When using the clustering algorithm on the reduced cases, the cases with 60, 30 and even 10 and 5 optimizations somewhat mirror the lower probability of *group 0* compared to the similar probabilities of *group 1* and *group 2* in the benchmark case. The cases with less optimizations either do not have any optimizations in *group 0* or assign the same probability to all three groups.

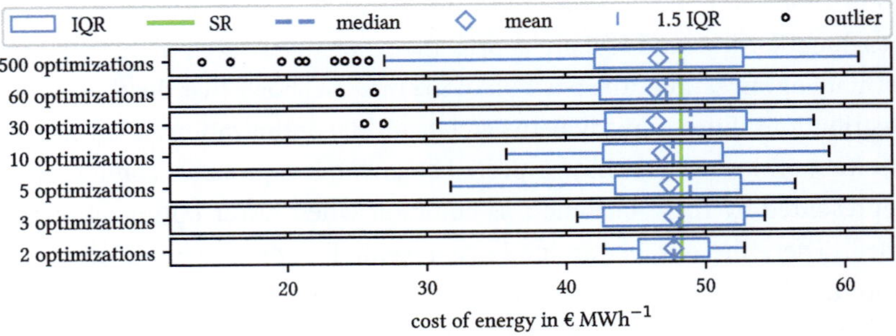

Figure 6.15: Comparison of optimization distribution result for different numbers of optimizations on the example of the cost of energy.

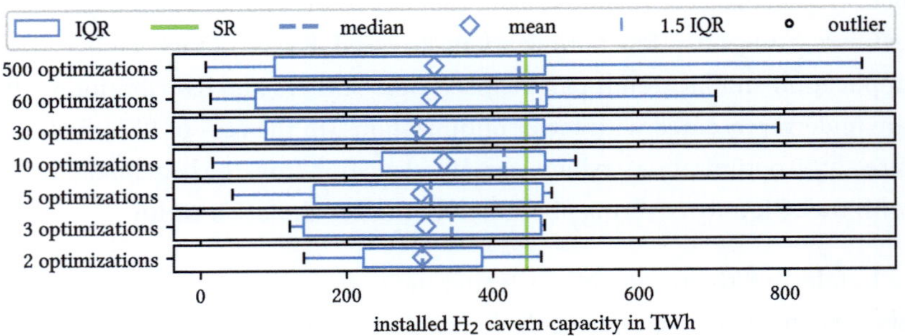

Figure 6.16: Comparison of optimization distribution result for different numbers of optimizations on the example of H_2 cavern storage.

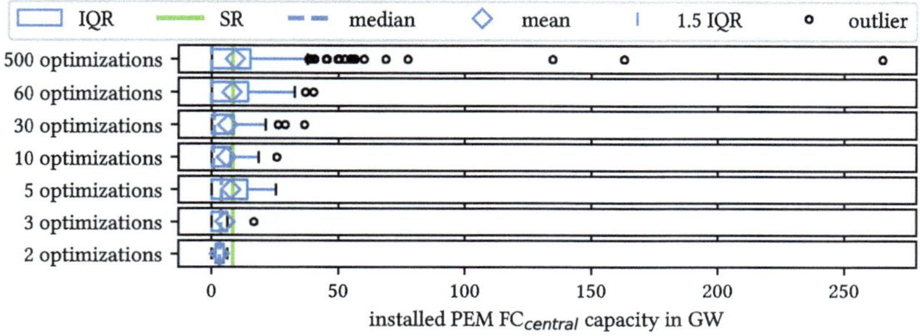

Figure 6.17: Comparison of optimization distribution result for different numbers of optimizations on the example of PEM $FC_{central}$.

Table 6.9: Probabilities of the benchmark clusters if results from stochastic optimization based on less model runs are assigned to the benchmark clusters in the case of 80 % emission reduction.

cluster	number of optimizations						
	500	60	30	10	5	3	2
group 0	0.11	0.13	0.17	0.00	0.20	0.33	0.00
group 1	0.42	0.43	0.50	0.40	0.40	0.33	0.50
group 2	0.47	0.43	0.33	0.60	0.40	0.33	0.50

6.4.2 Factor Effect based Meta Model

A meta model or surrogate model is a model that approximates the results of another model [244]. Meta models are used in many engineering and data science contexts and mainly serve the function of obtaining approximated results at a lower cost than obtaining exact values either from experiment or optimization. In the case of optimization relevant herein, a surrogate or meta model allows to calculate approximated results faster than re-running the optimization. A surrogate model created based on a reduced amount of data is useful if it lowers the number of optimizations required, i.e. the information extracted from each optimization run is increased.

One way to distinguish between different surrogate models is the distinction in black box and interpretable models. Black box models are generally models where the inner workings of the model are not known or too complex to be interpretable by the user, for example artificial neural networks. The use of many data driven models such as artificial neural networks depend heavily on the amount of input data and has in a previous attempt not been overly successful for a simplified energy system, as discontinuities are not represented well [95]. This behavior is known as a general obstacle for using artificial neural networks to approximate discontinuous functions [245]. The issue could be solved by applying techniques such as neutrons using a jump approximator [245] or by first applying a clustering in order to find the continuous sections and then approximating each cluster with an artificial neural network. As a general interpretability of the surrogate promotes a better understanding of the original system, herein the derivation of an interpretable meta model is attempted. Factor effect analysis, used to visualize the systems dependencies from the uncertain input parameters in section 6.3.1 and explained in detail in [103], is especially attractive as it allows for the determination of each technology costs effect on the respective output variable, i.e. the quality factor. As discernible in fig. 6.10 the dependency of the quality aspect in question, in this case the carbon emissions, from one of the factors, e.g. photovoltaic cost, can not be represented by a continuous function. This is limiting to the applicability of the factor effect methodology as a linear description model. Therefore instead of using the linear approximation of all higher order interactions only the basic factors, i.e. the technology costs as presented in fig. 6.10, are considered using kriging to approximate the quality aspect as a non-linear function of each factor.

Based on these considerations the approximation of the energy system model optimization output parameters inside the predefined parameter range of the probability distributions are investigated in order to reduce computational demand.

Meta model to Reduce Computational Demand

In preliminary optimizations the systems with higher allowed carbon emissions have been more difficult to predict with reduced optimization effort. Therefore, the meta model investigations are carried out based on the 80 % emission reduction boundary condition energy system. In order to reduce the computational demand while still adequately describing results, the goal of this investigation is reducing the number of necessary optimizations. To approximate the sometimes discontinuous slopes, a minimum of ten points is considered necessary, as less optimizations will lead to greater errors close to the discontinuities. Therefore a latin hypercube with ten factor settings has been chosen, that are adapted to the range of the lhc with 500 optimizations for each parameter, in order to achieve maximum information about the relevant parameter space. One advantage of the factor effect methodology is the use of mean variation over many input parameters, which makes it very robust, i.e. gives it a high prediction quality. As a lhc of 10 represents for each optimization only one possible combination, this advantage is lost. In order to retrain this advantage, the lhc is rotated, meaning that a second optimization is performed for each factor setting, with a different combination of the other parameters compared to the previous. Therefore for each factor setting now two results are available. This can be repeated several times in order to become more and more robust, i.e. independent of the single result. The price is the increased number of optimizations necessary, with 10 factor settings and six rotations a total of 60 optimizations becomes necessary. The different results for each factor setting can now be used as they are, retaining the information which input parameter combination led to which result. To further simplify, the mean of all the results at one factor setting can be calculated. A simple example of the procedure for three different factors is shown in fig. 6.18. The *mean calculation* yields the mean of the results for each factor setting independent of the other factors. With increasing number of rotations the *mean calculation* for each factor setting converges towards

145

Figure 6.18: Example for the methodology of rotation and mean calculation for a latin hypercube with two factors and three levels. The underlying equation is $result = 3\,factor_1 + factor_2^2$. Note that the use of the rotation enables six results with only three different settings for each factor.

	factor 1	factor 2	result
original:	0.167	0.833	1.194
	0.500	0.500	1.750
	0.833	0.167	2.528
rotated:	0.167	0.167	0.528
	0.833	0.500	2.750
	0.500	0.833	2.194
mean:	0.167		0.861
	0.500		1.972
	0.833		2.639
		0.167	1.528
		0.500	2.250
		0.833	1.694

the expected value of this result. That signifies that the average influence of the other factors is accounted for as well as the influence of the specific factor setting of the factor in question. An example calculation is included in fig. 6.18.

This rotation is herein performed up to elven times and the mean squared error (MSE) and description of variance (R^2) are calculated for the prediction of the results from the 500 optimizations. In some cases the used implementation of the kriging algorithm is unable to handle the matrices, leading to errors. Therefore the mean of all the results for one factor setting regarding a single factor is calculated as presented for three rotations in fig. 6.18. The resulting means are then fitted to the corresponding factor's settings, yielding an alternative description model. The results are classified according to table 6.10. A third option is explored by using the kriging fit when there are no errors and, when errors during fitting occur, switching to the approach of fitting to the mean. All three options are implemented, a comparison of the number of technologies belonging to each prediction grade according to table 6.10 is depicted in fig. 6.19.

Table 6.10: Quality grades of an output parameters meta model description.

	MSE	logical	R^2
grade 1	<5 %	V	>0.9
grade 2	5 - 15 %	V	0.6 - 0.9
grade 3	>15 %	V	≤ 0.6

An example of the changes in kriging fit performance between the two methodologies and varying number of rotations for onshore wind power and relative PV cost is shown in fig. 6.20. The rotated latin hypercube results for each factor setting are discernible, as well as their mean for three and six rotations and the five resulting kriging fits. For comparison the results of the 500 optimizations are included as well as a linear fit resulting from the data achived by applying the six fold rotated latin hypercube.

Between a factor setting of -0.75 and -0.5 the result depends on other factors, the trend of rising onshore wind installation is captured more adequately with higher rotation numbers. An advantage of the mean approach is the stability and continuous improvement with additional data. The advantage of kriging compared to using linear fits becomes clear by comparison of the fitting results: the discontinuity can be captured.

The above methodology is applied to the model. Three and six rotations, i.e. 30 and 60 optimizations in total, show a desirable increase in prediction quality in comparison with less optimizations. As the computational demand is lower than for the higher rotation settings, which show only minimal improvements in prediction quality, these settings are used. The use of kriging with prior

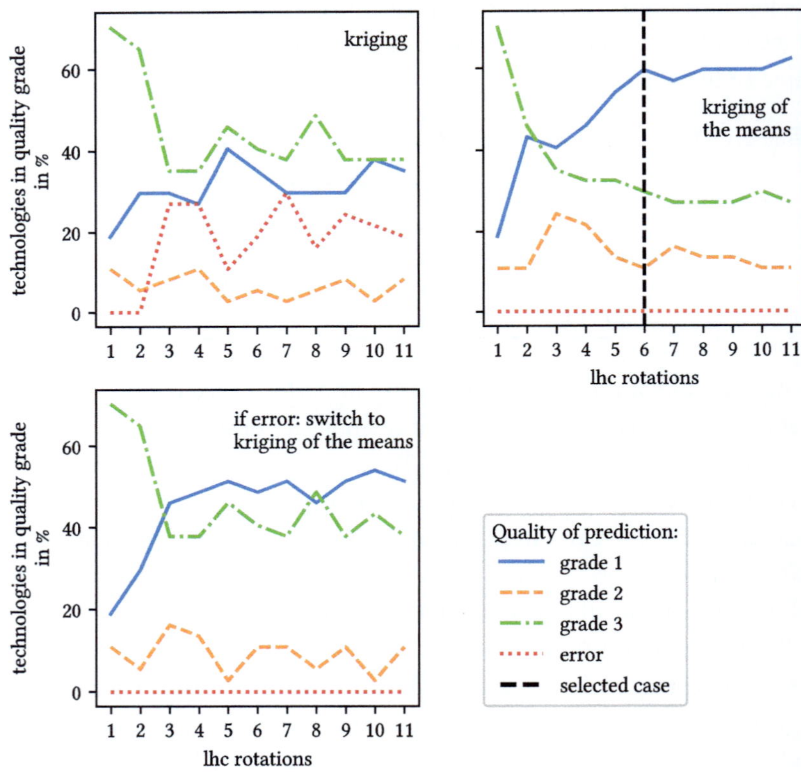

Figure 6.19: Change of the meta model quality for the prediction of installed capacities depending on the number of latin hypercube (lhc) rotations and the type of fitting used.

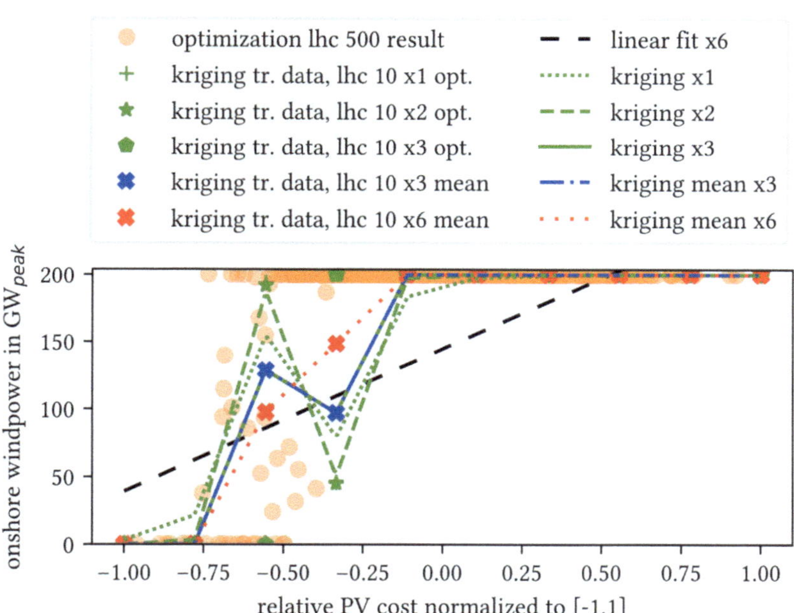

Figure 6.20: Comparison of the kriging fitting result for a one, two and three times rotated lhc (x1, x2 and x3) with each ten optimizations as well as the mean of three (x3) and six times (x6) rotated lhc on the example of onshore wind power plants benchmark optimization results that are to be described by the meta model inserted for reference. Kringing tr. is the training data used for kriging in the respective case.

mean calculation for each factor setting has a desirable performance regarding stability and result quality, therefore this option is applied. The grade 1 results, about 61% of the output parameters, are summarized in table 6.11, grade 2 results in table 6.12 and grade 3 results, which present 28% of the output parameters, in table 6.13.

Table 6.11: Parameters with a meta model performance grade 1. Res. refers to residential application. $|R^2|$ greater one results from numerical instabilities.

parameter	MSE	R^2
cost of electricity	1.49%	0.48
CO_2	8.73%	0.927
CC CH_4	2.87%	0.976
boilder med .CH_4	0.08%	0.798
CC H_2	1.03%	0.964
alk. electrolysis	3.47%	0.813
heating rod, old	0.50%	0.86
heating rod, new	0.07%	0.842
hp, new	0.08%	0.941
hp, central	0.04%	0.919
resistance heater	7.16%	0.904
electrode boiler	0.19%	0.246
direct heating, el.	0.13%	0.574
onshore wind	3.69%	0.332
offshore wind	0.87%	0.516
photovoltaic	2.70%	0.816
central heat storage	2.94%	0.878
liquid metal	2.08%	-2.318
molten salt	4.25%	-0.203
redox cap.	0.00%	-inf
res. heat storage, hp new	0.08%	0.941
res. heat storage, old	3.52%	0.635

Table 6.12: Parameters with a meta model performance grade 2.

parameter	MSE	R^2
cond. boiler, old	22.00%	0.761
cond. Boiler, new	64.62%	0.73
battery cap.	4364%	0.731
H_2 cavern	10.19%	0.758

Table 6.13: Parameters with a meta model performance grade 3. $|R^2|$ greater one results from numerical instabilities.

parameter	MSE	R^2
SC CH_4	13777539%	-1593
CH_4 cavern	945%	-1.48
boiler	710.9%	0.11
SC H_2	133.4%	0.65
PEM $FC_{central}$	258.1%	0.40
methanation	1055%	-2.01
PEM $FC_{delivery}$	8984%	-1.77
boiler med. H_2	8588%	-1.46
direct heating, H_2	24.44%	0.31
H_2 on-site	6487%	0.14
res. heat storage, new	50.42%	-0.42

According to the distinction in table 6.10 grade 1 results are desirable as they offer either an acceptable mean squared error or a model with a high coefficient of determination.

6.4.3 Comparison of Benchmark Results With Reduced Computational Demand Approaches

To be able to judge the possibility to use either a distribution with less optimizations or the above presented meta models, the results of both approaches are compared with the benchmark, i.e. the stochastic optimization of the 80 % emission reduction case based on 500 optimizations. In order to enable potential users to choose which of the approaches is more appropriate under which circumstances, the comparison is performed firstly for the box-plot presentation as used in section 6.2 to communicate the stochastic optimization results and secondly with respect to the possibility of clustering application to gain insight into probable system configuration clusters. As from the meta models two configurations are likely to have a desirable trade-off between the number of optimizations and output quality the comparison is performed for the meta models based on three and six rotations of the latin hyper-cube (MM-x3, MM-x6), as well as two stochastic optimizations with 30 and 60 optimizations each.

Comparison Regarding Box-Plots

The comparison is performed on examples based on a selection from the different quality grades depicted in tables 6.11 to 6.13. Figure 6.21 shows the comparison between benchmark and meta model results for a grade 1 parameter. Figure 6.22 is an example of a grade 2 result quality and fig. 6.23 of a grade 3 result. For comparison the results of stochastic optimizations with the same number of optimizations needed to construct the meta models, i.e. 30 and 60, are displayed as well.

The comparison for cost of energy depicted in fig. 6.21 shows small deviations from the benchmark box plot. Result range as well as the approximate position of the mean are comparable. For the x3-meta model and the stochastic optimization with 30 optimizations the distribution is shifted slightly to the higher costs compared to the benchmark. In the case of the x3-meta model mean and median lie closer together, while in the reduced case with 30 optimizations they are further apart compared to the benchmark. The meta models are able to predict the presence of outliers and their approximate position, while in the reduced stochastic cases the distribution is narrower with less outliers. The x6-meta model approximates position of mean and median, whereas

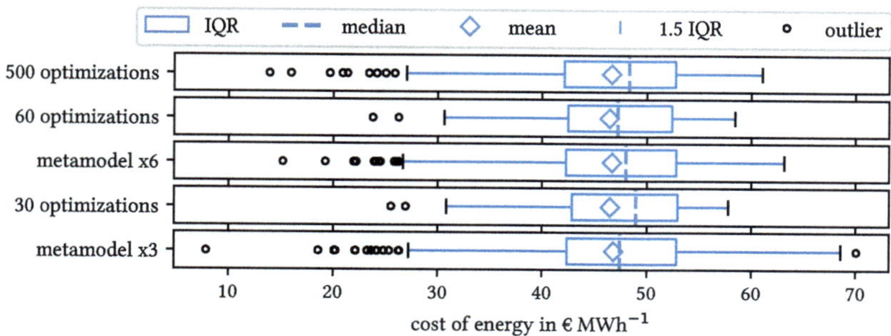

Figure 6.21: Grade 1 comparison of metamodel approximation and 500 optimization distribution result benchmark on the example of the cost of energy.

in the 60 optimization reduced case these are closer together, indicating a different distribution shape.

In the grade 2 case, hydrogen cavern storage, position and width of the distribution are similar between benchmark and meta models which do not reach the quality of the reduced cases. The IQR and mean of the meta models, as well as the 30 optimization reduced case lie closer together, indicating that the shape of the distribution is approximated more accurately by the stochastic optimization with 60, data points. The distribution predicted by the meta models is shifted to higher capacities, with the x3-meta model overestimating the benchmark distribution more than the x6-meta model. The reduced optimization cases show narrower distributions than the benchmark and meta model predictions.

Grade 3 example, PEM FC for central application, is depicted in fig. 6.23. The x6-meta model underestimates the width of the IQR and the mean result by about 50 % compared to the benchmark case and the case of 60 optimizations. Again the reduced stochastic distributions approximate the benchmark more closely than the meta models regarding distribution width and position of the mean. None of the reduced computational demand alternatives is able to

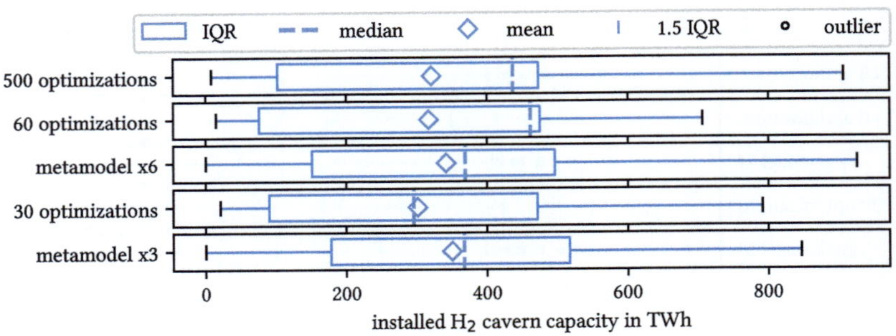

Figure 6.22: Grade 2 comparison of metamodel approximation and 500 optimization distribution result benchmark on the example of hydrogen cavern storage .

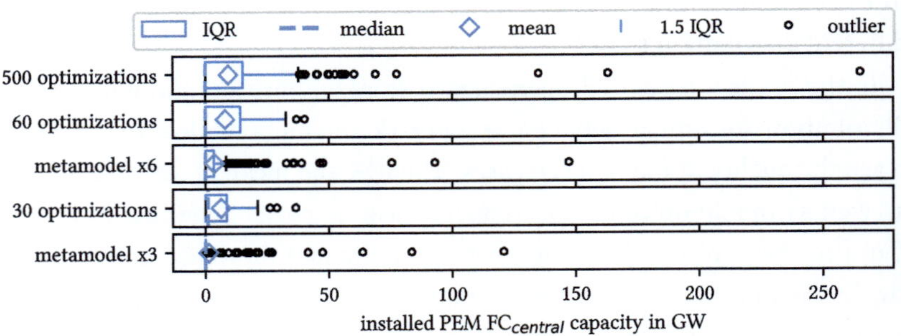

Figure 6.23: Grade 3 comparison of metamodel approximation and 500 optimization distribution result benchmark on the example of PEM FC in central application.

predict the maximum amount of possible optimal PEM implementation above 250 GW. The meta models yield cases at about 150 and 120 GW respectively, while the 30 and 60 optimization based stochastic optimizations only predict a maximum of about 40 GW.

The comparison of the two different meta models, x3 and x6, with 30 and 60 optimizations used for training respectively, with the result of the stochastic investigation based on 500 optimizations and the results from stochastic optimization with 30 and 60 optimizations is presented regarding the means in tables 6.14 and 6.15 and regarding the IQR in tables 6.16 and 6.17.

Table 6.14: Comparison of scenario, stochastic and meta model result regarding the mean - part 1.

technology	sc.	80 % red. stochastic 500 opt	30 opt	60 opt	meta model MMx3	MMx6	unit in
onshore wind	200	188	181	183	159	177	GW_{peak}
offshore wind	85.0	83.4	82.2	85.0	89.8	82.5	GW_{peak}
PV	2.08	1.76	1.72	1.74	1.81	1.74	TW_{peak}
battery	0	1.43	0	0	8.40	3.78	GWh
redox	0	0	0	0	0.35	0.18	GWh
PEM EC central	8.37	9.39	6.4	8.09	1.35	3.70	GW
PEM EC on site	0	0.08	0.14	0.06	0.41	0.47	GW
AEL	1.19	0.97	0.95	0.96	0.98	0.93	TW
methanation	0	0.75	0.79	0.87	0.21	0.88	GW
CH_4 cavern	0	3.30	3.44	3.80	1.07	3.43	TWh
H_2 cavern	447	322	302	318	351	343	TWh
H_2 on site	0	0.72	0.64	0.33	1.46	1.06	GWh
SCGT CH_4	0	0	0	0	0.10	0.32	GW
CCGT CH_4	11.0	65.13	76.7	67.2	68.2	64.4	GW
SCGT H_2	0	3.46	4.14	3.6	3.45	4.01	GW
CCGT H_2	151	112	103	110	110	110	GW

The comparison herein comprises only the x6-MM and the 60 optimizations reduced case, because acceptable trade off between accuracy and computational effort are very dependent on model and user requirements. For some purposes the 30 optimizations reduced case, which for many technologies is comparable to the x6-MM, or the x3-MM which is less accurate, might suffice. The 60 optimizations reduced case as well as the x6-MM generally

Table 6.15: Comparison of scenario, stochastic and meta model result regarding the mean - part 2.

| | | | 80 % red. | | | | unit |
| | | | stochastic | | meta model | | |
technology	sc.	500 opt	30 opt	60 opt	MMx3	MMx6	in
heating rod, old	348	289	281	284	275	276	GW
heating rod, new	56.7	55.0	54.8	54.5	53.2	54.9	GW
HP, new	11.8	11.3	11.2	11.2	11.9	11.2	GW
central HP	21.4	22.0	22.1	22.0	22.9	22.1	GW
resistance heater	15.2	9.66	8.49	9.45	7.13	8.67	GW
electrode boiler	19.3	20.4	20.3	20.8	20.1	20.2	GW
direct el. heating	101	97.9	97.5	98.1	103	98.3	GW
boiler	0	0.07	0.13	0.09	0.06	0.07	GW
boiler mid CH_4	8.69	8.39	8.37	8.34	8.89	8.50	GW
cond. boiler, old	9.15	4.13	4.36	3.76	4.43	4.13	GW
cond. boiler, new	0	0.76	0.86	0.75	0.63	0.63	GW
boiler mid H_2	0	0.02	0.04	0.02	0.15	0.13	GW
direct heating H_2	10.3	14.2	15.2	14.3	18.1	17.8	GW
central hs	376	609	654	606	612	571	GWh
liquid metal hs	532	521	521	522	607	490	GWh
molten salt hs	105	118	120	121	110	116	GWh
resid. hs, new	15.6	8.3	7.38	7.83	5.47	5.11	GWh
resid. hs HP, new	41	39	38.9	39.1	41.5	38.9	GWh
resid. hs, old	1006	762	724	750	647	683	GWh
cost of energy	48.3	46.7	46.5	46.5	46.8	46.7	€ MWh^{-1}
CO_2	10.5	58.6	67.2	61.1	60.0	60.5	Mtons

Table 6.16: Comparison of stochastic and meta model result regarding the IQR - part 1.

	80 % red.										
	stochastic						meta model				
	500 opt.		30 opt.		60 opt.		MMx3		MMx6		unit
technology	25 %	75 %	25 %	75 %	25 %	75 %	25 %	75 %	25 %	75 %	in
onshore wind	200	200	200	200	200	200	133	189	165	203	GW_{peak}
offshore wind	85	85	85	85	85	85	87.7	93.7	78.9	87.8	GW_{peak}
PV	1.12	2.21	1.08	2.21	1.05	2.21	1.19	2.39	1.14	2.22	TW_{peak}
battery	0	0	0	0	0	0	0	13	0	0	GWh
redox	0	0	0	0	0	0	0	0	0	0	GWh
PEM EC central	0	15	0	8.68	0	14.3	0	0	0	3	GW
PEM EC on site	0	0	0	0	0	0	0	0.66	0	0.73	GW
AEL	0.58	1.25	0.56	1.25	0.54	1.26	0.58	1.34	0.54	1.22	TW
methanation	0	1	0	1.11	0	1.24	0	0	0	0	GW
CH_4 cavern	0	3.76	0	5.22	0	6.44	0	0	0	1.49	TWh
H_2 cavern	103	473	90.8	472	76.8	475	178	518	151	497	TWh
H_2 on site	0	0	0	0	0	0	0	0	0	0	GWh
SCGT CH_4	0	0	0	0	0	0	0	0	0	0	GW
CCGT CH_4	0	139	0	143.4	0	148	7.31	124	0.61	125	GW
SCGT H_2	0	4	0	4.58	0	4.82	0	5	1.09	4.89	GW
CCGT H_2	51	164	47.8	164	43.6	164	60	162	56	159	GW

represents the mean of the stochastic optimization well and both exhibit an advantage compared to the scenario approach. For some technologies with low implementations, in the example herein battery, redox flow battery, PEM EC on-site, hydrogen on-site storage and the medium temperature hydrogen boiler, the stochastic result is overestimated by the meta models, while the reduced stochastic cases still approximate reasonably well. To the mind of the author the approximation observed in case of the meta models is still preferable to the scenario result. The scenario approach, which suggests zero installed capacity for these technologies, is closer to the stochastic result in terms of absolute installed capacity. Yet overestimating a small contribution in the energy system compared to the implications of suggesting a technology has no role in an optimized future energy system seems preferable to the author.

Table 6.17: Comparison of stochastic and meta model result regarding the IQR - part 2.

					80 % red.						
		stochastic				meta model					
	500 opt.		30 opt		60 opt		MMx3		MMx6		unit
technology	25 %	75 %	25 %	75 %	25 %	75 %	25 %	75 %	25 %	75 %	in
heating rod, old	251	336	241	331	236	332	246	306	244	314	GW
heating rod, new	53.0	56.3	52.9	55.5	53.0	55.2	51.0	53.9	52.9	55.3	GW
HP, new	11.0	12.0	11.0	12.0	11.0	12.0	11.6	12.8	10.9	11.9	GW
central HP	20.6	23.7	20.6	23.7	20.6	23.8	21.7	24.2	21.0	23.5	GW
resistance heater	0	16.8	0	17.0	0	17.3	0	12.9	0.02	15.77	GW
electrode boiler	19.5	21.4	19.6	21.1	20.0	21.6	19.8	20.6	19.9	20.7	GW
direct el. heating	96.0	101	95.4	101	95.7	101	100	107	97.3	99.9	GW
boiler	0	0	0	0.05	0	0	0	0.10	0	0.12	GW
boiler mid CH_4	8.13	8.69	8.08	8.69	8.13	8.69	8.65	9.30	8.24	8.84	GW
cond. boiler, old	0	7.83	0	7.56	0	6.31	1.99	6.86	1.17	7.02	GW
cond. boiler, new	0	1.66	0	1.71	0	1.55	0	1.09	0	1.19	GW
boiler mid H_2	0	0	0	0	0	0	0.02	0.24	0	0.22	GW
direct heating H_2	7.68	22.2	8.77	21.9	9.16	22.4	14.5	21.9	15.7	20.5	GW
central hs	368	926	406	920	381	983	391	813	350	788	GWh
liquid metal hs	500	536	485	544	500	536	554	655	428	533	GWh
molten salt hs	112	116	111	117	112	116	78.7	134	87.9	130	GWh
resid. hs, new	3.84	13.3	2.75	11.8	3.8	11.6	4.80	6.20	4.76	5.50	GWh
resid. hs HP, new	38.3	41.7	38.35	41.7	38.2	41.7	40.5	44.6	38.1	41.4	GWh
resid. hs, old	517	986	507	960	496	981	564	734	574	801	GWh
cost of energy	42.1	52.8	42.9	53.0	42.5	52.4	42.4	52.9	42.3	52.8	€ MWh^{-1}
CO_2	0	122	0	127	0	131	4.12	101	6.65	111	Mtons

Only in the case of PEM EC in central application the meta model underestimates the cost optimal installed mean by about 60 %. The expected value, i.e. the mean of all other technologies as well as of carbon emissions and system cost, is predicted well. The system cost, respectively cost of energy, estimates the stochastic result almost exactly in all cases with reduced computational demand. Therefore these approaches account for the lower expected system cost compared to the stochastic optimization result due to Jensen's inequality correctly.

Regarding the IQRs the value of the upper quartile is strongly underestimated by the meta models in the case of methane cavern storage, PEM EC central and residential heat storage in new buildings. While in the latter case stochastic optimization with 30 and 60 approximates the benchmark distribution reasonably well, the upper end of the IQR is predicted to be about twice that of the benchmark in the case of methane storage. The IQR of PEM EC in central application is estimated well with 60 optimizations, while 30 underestimate the upper end by over 40 %. For direct hydrogen heating the lower quartile is overestimated by the x6-MM with 15.7 instead of 7.7 GW. The potential for the reduction of carbon emissions in a cost optimal system are underestimated with an IQR from 6.6 to 111 Mtons instead of 0 to 122 Mtons in the stochastic result. The reduced stochastic cases with 30 and 60 optimizations are able to predict this range more accurately by resulting in 0 to 127 and 0 to 131 Mtons of annual emissions.

Summarizing all approaches present additional value compared to the scenario result, while the direct use of stochastic optimization with less optimizations yields more accurate results regarding IQR and mean compared to the meta models based on the same number of optimizations.

Comparison Regarding Cluster Generation

The clustering algorithm defines the clusters based on the available input data. Therefore clusters can and will usually change with each additional input data set. To allow for a comparison with the benchmark the number of clusters is set to three by iteratively raising the minimum cluster size until three clusters emerge, which is the number of clusters resulting from the benchmark clustering for the 80 % emission reduction case. The clusters from the reduced computation approaches most similar to each of those from the benchmark stochastic optimization are assigned to the same group number for comparison. Clustering is performed for the two compared meta models, which are applied to all technology outputs from the optimization to achieve data sets comparable to the optimization results, this includes sub-results like the maximum energy flow in the gas grid as well as from and to caverns, as well as the installed battery and redox flow battery power in addition to their capacity. The optimization results from stochastic optimization with 30 and 60 runs contain all the necessary information and are clustered as well for comparison. Each technology's mean is calculated for each cluster and methodology. To allow a quick overview regarding model quality, histograms comparing the performance of the relative difference between reduced computational approach and benchmark mean in each cluster are computed. All results greater 1.1, i.e. that overestimate the result by more than 110 % are grouped at 1.1 to allow for representation. The results of the reduced computational demand approaches that use 60 optimizations are displayed in fig. 6.24, the results from 30 optimizations are depicted in fig. C.27.

The technology-wise comparison answers how similar the clusters resulting from the clustering of the reduced computational demand approaches are to those of the benchmark. The higher the number of technologies with a deviation close to zero, the more similar the clusters. The histograms show that most of the technologies' means inside the respective clusters are within 10 % relative deviation from the mean of the respective technology of the benchmark cluster. The stochastic optimization with 60 optimizations yields small deviations for more technologies than the MMx6. The cases in which the MMx6 model overestimates the benchmark severely are summarized at 1.1 and make up about 20 % of the technologies. These are mainly the technologies which are installed very little or not at all in the benchmark case. The opposite effect is visible regarding the results from the 60 optimizations with a peak at -1. Here the smaller number of cases causes, that a low probability of installation is not part of the results at all, meaning the technology is underrepresented in the clusters compared to the results from 500 optimizations. An additional difference is discernible regarding the probability with which the different clusters are predicted. While the 60 optimizations predict the occurrence of the clusters relatively accurately with a maximum of four percent points deviation, the MMx6 model deviates up to 18 percent points in case of the probability for group 3 to be optimal. Results for the case of 30 optimizations are similar as can be seen in fig. C.27 entailed in appendix C.5. Based on these findings the clustering of the reduced computational demand approaches

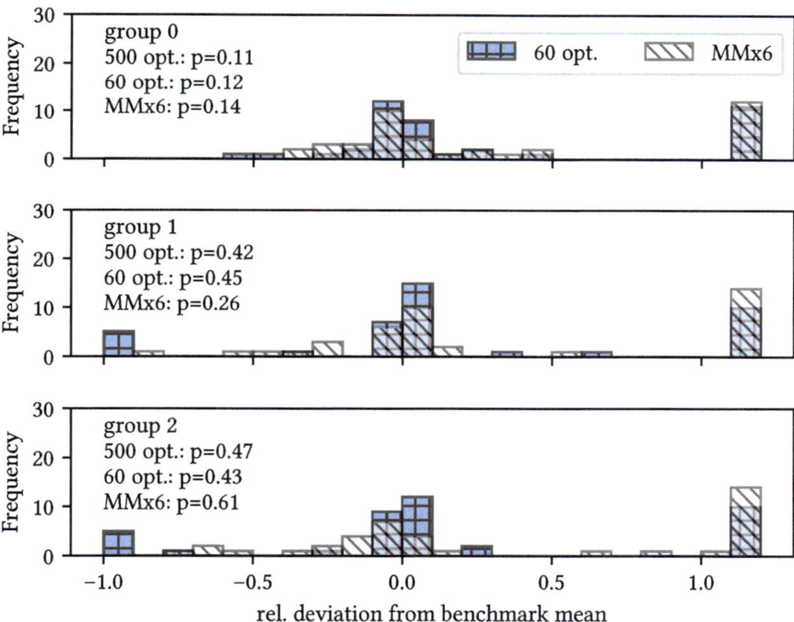

Figure 6.24: Histogram comparing the reduced computational demand approaches regarding the relative difference from the mean of installed capacity of each technology to the benchmark inside of the clusters. Reduced computational demand at 60 optimizations.

yields clusters similar to those generated by the full number of optimizations in terms of technology mean as well as in terms of cluster probability. An important limitation is that the number of clusters must be known beforehand in order to achieve comparable clusters.

6.4.4 Summary of the Comparison of Benchmark Results with Reduced Computational Demand Approaches

Generally the use of less optimizations is feasible if a certain loss of information is acceptable. The use of 60 optimizations and, depending on the degree to which information loss is justifiable, also 30 optimizations yield enough data to approximate means and IQRs of the output distributions. The use of a meta model with 60 optimizations for training compared to directly using 60 lhc distributed optimizations has advantages regarding the prediction of less probable cases of very high or very low use of a technology. For many technologies the prediction by the meta model is of high quality, with low MSE and a high ability to describe variance. For some technologies, generally, but not exclusively, technologies with little to no implementation in the resulting energy systems, the results from the meta model can not be considered an acceptable approximation. As the meta model training data is evenly distributed in the parameter space which is considered reasonable or possible, the input parameter distribution over which the meta model is evaluated can be changed in post processing. Therefore the metamodel is advantageous to judge the implementation of the most important technologies under different input parameter distribution assumptions, but it can not answer if a niche technology will be important, as results could always be artifacts. The metamodel offers more flexibility to changing input assumptions. The use of input parameter distributions with only 60 data sets yields comparable results for the specific case, but has a high data quality as each system that makes up the final distributions is actually an optimized system. A disadvantage of this approach is that unlikely systems are underrepresented due to the low number of optimizations and that a change in input parameter distribution assumptions warrants a re-run of all, in this case 60, optimizations. The advantage of the metamodel regarding its flexibility is lost if an input parameter that has not been considered as uncertain changes. The use of clustering to analyze the results achieved with reduced computational demand is generally possible. Cluster probability and its structure as determined by the dominant technologies are approximated. It is important to note that for this it is necessary to either define a cluster to which the results can then be assigned with an acceptable precision, or to at least give the number of clusters in order to obtain comparable results to the case of 500 optimizations. Clustering without these provisions will still yield results which help interpret the generated data.

Chapter 7

Discussion

In this chapter results of the optimizations and the different methodologies to interpret them are discussed regarding usability, interpretability and information gain. In order to do so first the results from the scenario approach and those of the stochastic optimizations are compared. Based on the comparison advantages, drawbacks and possible limitations regarding information gain and usability are discussed and a closer look is taken to the interpretation of the stochastic optimization with the use of factor effects and cluster analysis. Finally the use of the derived factor effect based meta models to limit the necessary number of optimizations is discussed in comparison to the use of stochastic optimization based on the same number of optimizations.

7.1 Comparison of stochastic and scenario results

In order to compare the planning value of both, scenario and stochastic approach, mean and scenario result for the different technologies in all three emission reduction cases are confronted in tables 7.1 and 7.2, first of which covers electricity generation and gas storage technologies whereas the second involves heating technologies and CO_2 emissions as well as cost of energy. One evident difference between scenario and stochastic results is the implementation of battery and redox electricity storage. While in the scenario battery is not implemented at all, the mean of the stochastic optimization lies between 1.43 and 3.15 GWh. Similarly, stochastic optimization yields mean redox storage capacities between 0.82 and 2.11 GWh for the 95 % and 100 % emission reduction cases with stochastic optimization, compared to no installed capacity in the respective mean scenarios. As expected from Jensen's inequality, the expected cost of energy is lower in the stochastic cases than that calculated in the scenario. In the optimized cases this difference is greater at greater allowed carbon emissions. With an overestimation of 3.5 % this effect is limited. Another notable difference between stochastic and scenario results concerns the optimal amount of reversible PEM fuel cells. While in the 100 % emission reduction scenario optimization the optimal amount is 7.66 GW, the corresponding stochastic optimization yields, with a mean of about 31 GW, about four times as much installed capacity. The third main difference is related to heating technologies: optimal direct heating with hydrogen covers about twice as much capacity (15.7 GW) according to the stochastic optimization compared to the scenario approach (8.5 GW for 100 % emission reduction). Other notable differences between scenario results and stochastic optimization concern the cases where the use of fossil methane is allowed:

1. Methanation: In the case of a non entirely renewable energy system the scenario results in zero installed capacity, while in the stochastic cases a mean of 0.75 GW are implemented.

2. CH_4 cavern: In accordance with the methanation implementation the stochastic result leads to 3.3 TWh instead of 0 TWh storage capacity in the non entirely renewable cases.

3. CCGT CH_4: Also here the stochastic optimization reveals the implementation of 29.9 GW in the 95% and 65.1 GW in the 80 % emission reduction case while in the scenarios the optimally installed capacity is 11 GW.

4. SCGT H_2: While in all the scenario cases this technology is not implemented, the stochastic optimization yields a mean of about 3.46 GW installed capacity.

5. Condensing boiler, old: In the stochastic case a mean of 2.47 GW are installed in the renewable case, 11.7 GW in the 95 % reduction case and 4.13 GW in the 80 % reduction case, while in the corresponding scenario optimizations the result is zero in the renewable case and 9.15 GW in the non-renewable cases.

6. Condensing boiler, new: In new houses the stochastic optimization leads to a mean of 0.03 and 0.76 GW condensing boilers installed in houses in the 95 % and 80 % emission reduction case, while for the scenario approaches the result is zero.

7. CO_2 emissions: The scenario optimization suggests a lower amount of emitted CO_2 under cost optimal conditions than the stochastic approach. In numbers the scenario yields 10.47 million tons and the stochastic result leads to a mean of 24.9 million tons in the 95 % reduction assumption and 58.52 million tons in the 80 % emission reduction scenario.

Neither the scenario approach nor the corresponding results contain information regarding future input parameter distributions and correlations between these. Nevertheless the scenario result represents, regarding most technologies, a good guess of the mean of the probable system regarding the technology implementations. Yet the analysis reveals some significant deviations between scenario result and stochastic optimization result mean, e.g. regarding probable CO_2 emissions and the implementation of certain technologies. Also technologies with a lower probability of being implemented at all will many times not be included in the scenario result or are, if included, overrepresented as their overall probability of application is low.

The box-plots used to visualize the stochastic optimization results show that for most technologies the IQR is comparably big, many times in the or-

Table 7.1: Comparison of scenario results with the mean, respectively expected value of the stochastic optimizations for electricity generation and gas storage technologies. Red. refers to the emission reduction scenario.

| technology | 100 % red. | | 95 % red. | | 80 % red. | | unit |
	sc.	stoch.	sc.	stoch.	sc.	stoch.	in
onshore wind	200	188	200	188	200	188.2	GW_{peak}
offshore wind	85.0	83.4	85.0	83.4	85.0	83.4	GW_{peak}
PV	2.19	2.24	2.08	2.02	2.08	1.76	TW_{peak}
battery	0	3.15	0	2.49	0	1.43	GWh
redox	0	2.11	0	0.82	0	0	GWh
PEM EC central	7.66	31.38	8.37	19.3	8.37	9.39	GW
PEM EC on site	0	0.06	0	0.06	0	0.08	GW
AEL	1.26	1.26	1.19	1.13	1.19	0.97	TW
methanation	3.62	3.44	0	0.75	0	0.75	GW
CH_4 cavern	15.5	14.9	0	3.30	0	3.30	TWh
H_2 cavern	467	486	447	413	447	322	TWh
H_2 on site	0	0.26	0	0.26	0	0.72	GWh
SCGT CH_4	0	0	0	0	0	0	GW
CCGT CH_4	0	0	11.0	29.9	11.0	65.13	GW
GE CH_4 on site	0	0	0	0	0	0	GW
GE CH_4 central	0	0	0	0	0	0	GW
SCGT H_2	0	3.48	0	3.45	0	3.46	GW
CCGT H_2	163	165	151	133	151	111.65	GW
GE H_2 central	0	0	0	0	0	0	GW
PEM FC central	7.7	31.3	8.4	19.3	8.4	9.39	GW
GE H_2 on site	0	0	0	0	0	0	GW
PEM FC on site	0	0.06	0	0.06	0	0.08	GW

Table 7.2: Comparison of scenario results with the mean, respectively expected value of the stochastic optimizations for heating technologies, cost of energy and CO_2 emissions.

technology	100 % red. sc.	100 % red. stoch.	95 % red. sc.	95 % red. stoch.	80 % red. sc.	80 % red. stoch.	unit in
heating rod, old	355	353	348	329	348	289	GW
heating rod, new	57.3	57.9	56.7	55.2	56.7	55.0	GW
HP, old	0	0	0	0	0	0	GW
HP, new	11.8	11.6	11.8	11.5	11.8	11.3	GW
central HP	21.0	21.0	21.4	21.8	21.4	22.0	GW
resistance heater	14.4	14.4	15.2	12.0	15.2	9.66	GW
electrode boiler	21.6	22.0	19.3	20.4	19.3	20.4	GW
direct el. heating	100	99	101	99.2	101	97.9	GW
boiler	0	0.02	0	0.02	0	0.07	GW
boiler mid CH_4	8.69	8.24	8.69	8.39	8.69	8.39	GW
direct heating CH_4	0	0	0	0	0	0	GW
cond. boiler, old	0	2.47	9.15	11.7	9.15	4.13	GW
cond. boiler, new	0	0	0	0.03	0	0.76	GW
boiler mid H_2	0	0.12	0	0.02	0	0.02	GW
direct heating H_2	15.7	8.5	10.3	8.2	10.3	14.2	GW
central hs	371	382	376	412	376	609	GWh
liquid metal hs	524	517	532	515	532	521	GWh
molten salt hs	112	116	105	111	105	118	GWh
resid. hs, new	16.0	16.4	15.6	12.7	15.6	8.3	GWh
resid. hs HP, new	41	40	41	40	41	39	GWh
resid. hs, old	1060	1044	1006	919	1006	762	GWh
resid. hs HP, old	0	0	0	0	0	0	GWh
cost of energy	48.42	48.22	48.27	47.42	48.27	46.66	€ MWh^{-1}
CO_2	0	0	10.47	24.89	10.47	58.58	Mtons

der of magnitude of the technologies mean implementation and the scenario result. This variability is not accounted for in case of the scenario analysis. An additional drawback of the scenario approach is that there is no information as to the different probable or possible system designs depending on the input parameter distributions. Cluster analysis applied to the stochastic optimization results reveals different system layouts, especially important for the non completely renewable boundary condition. The scenario result can always only be one representation of one of the clusters of possible systems. In the herein analyzed cases the scenario result represents the following: For 100 % carbon emission reduction the group 1 with 93.2 % probability, for 95 % emission reduction in group 3 with 21 % probability and for the 80 % emission reduction group 2 with 46.8 % probability. As a single representative of these clusters the scenario result visualizes one possible system layout, yet also here the variability of each technology's implementation, represented by the technologies IQR in the cluster, is not accounted for. Another possible information achieved by the use of cluster analysis and therefore not attainable by the use of the scenario approach is the influence that the uncertain technology costs have on the optimal system layout. This allows for interpreting observed changes in cost trends as to how probable which cluster describing a certain system layout becomes. Factor effect analysis applied to the scenario results allows to quantify the influence one technology's cost has on any technology's implementation. A part of this information can be achieved by conventional sensitivity analysis, yet there is no guarantee that the most relevant influences can be discovered that way and no knowledge if important effects have been overlooked.

The main advantage of the scenario result is its comprehensibility. Yet trust in results is always diminished by the fact that there is no inherent sensitivity analysis and thus no information as to the results robustness is available. Additionally the interactions between different parameters are not accounted for, even if traditional sensitivity analysis is performed. To the mind of the author the biggest disadvantage of the scenario result is the seeming certainty, that is prone to deceive even experts as the high dimensionality and complex interactions do not allow for an a priori estimation of influences and uncertainties. Recent publications show recognition of this challenge by researchers, as there has been significant investigation in robust optimization applied to energy systems optimization. Stochastic optimization as performed herein is inherently robust, as reasonable variations of the assumed input parameter distributions will only have little effect on the output distributions, since a great part of the parameter space is still identical to the prior configuration.

Regarding the role of energy systems optimizations in the transition towards a renewable energy based energy system necessary cost will always be overestimated by scenario analysis due to Jensen's inequality, even if this overestimation is small in the herein analyzed cases and the meaningfulness of optimized system costs are disputable. Probable cost optimal carbon emissions are greatly underestimated by a factor of about 2.5 in the case of 95 % carbon emission reduction and a factor of almost six in the case of 80 % carbon emission reduction. This suggests that, especially to evaluate the impact of possible emission aims for planning of the energy transition, scenario analysis might not be sufficient. This is supported by the different possible system layouts, especially for the not completely renewable energy futures, that can not be represented by scenario analysis and could lead to non optimal policy decisions by neglecting obtainable information that represents a large portion of possible optimal outcomes.

7.2 Stochastic optimization with reduced computational effort

The additional information gain achieved by stochastic optimization compared to scenario analysis comes at the cost of additional computational effort. As the factor effects capture the dependency of the systems output parameters on the uncertain inputs, a description model or meta model based on these can be used to predict optimization results. Another possibility is to directly use less optimizations in stochastic optimization.

The technologies distribution resulting from the stochastic optimization can be estimated by the meta model based on factor effects utilized herein for most technologies. The surrogate model results allow the identification of technologies with big IQRs where a high uncertainty regarding their role in a cost optimized future energy system is present. Also the technologies which, under the assumptions made, will play an important role in all or the majority of probable system configurations can be assessed. An important limitation of the meta model is that for some of the technologies the prediction quality is rather poor. The use of the meta model regarding computational effort and result quality does not show advantages over the use of the same number of optimizations used directly stochastic optimization. Stochastic optimization with 30 or 60 runs yields box-plots comparable to 500 runs, which represents the benchmark case. One drawback of using less optimizations is that extreme cases with a low probability, i.e. few optimizations showing the specific result in the benchmark case, are many times not part of the results at all. This is one of the aspects where a meta model could prove useful, as these, even though the extreme cases are not predicted accurately, at least account for them. Another possible application of meta models is the option to change the probability distributions after the optimization has taken place, if training data is chosen that covers the required parameter space.

Chapter 8

Summary and Conclusion

ESO models are a valuable tool and used widely as a basis for decision making in public and private contexts. As most of the ESO models concern cost optimal system layouts under expected future developments, uncertainty is present. The value of ESOs is often limited because users, i.e. readers of the reports, do not agree completely with the author on the specific assumptions made, especially regarding expected future costs of technologies. The nature of uncertainty and the complexity of the cause effect relations in ESO models make it difficult or impossible even for experts to asses the impacts of uncertain input data on ESO results without detailed investigation of the uncertainty.

8.1 Summary of the Influence of Uncertainty

Based on a simple renewable energy system an experimental investigation of the effect of Jensen's inequality on the optimization results and especially the optimization objective has been performed. Regarding the optimization objective, in this case the system cost, the systematic overestimation predicted by Jensen's inequality has been confirmed. This holds for applications in which the mean scenario, i.e. a scenario based on each uncertain quantities mean is used, compared to a stochastic optimization with a discretization of the uncertainty. The effect has been shown to increase with increasing number of uncertain parameters and increasing probability distribution width. For the model of the German energy system herein investigated the overestimation of necessary cost is about 3.5 % for the case of 80 % reduction in carbon emission and 0.4 % for a renewable system. Regarding the optimized technologies capacity it has been shown that even small uncertainties can lead to comparably big variations in the optimal installed capacity. Also here generally greater uncertainties and a greater number of uncertain parameters leads to a broader distribution in the results and therefore a higher uncertainty.

8.2 Summary Energy System Optimization under Uncertainty

Based on a relatively complex ESO model that takes into account the greatest parts of electricity, heating and transport sector in Germany, the effects of uncertainty on ESO have been investigated for a more realistic, i.e. complex example.

One important property of stochastic optimization is its inherent robustness regarding variations of the input parameter distributions. Where a scenario investigation can yield completely different results if the input parameters are changed a little, variations of the distributions in the case of stochastic optimization have a small effect on the resulting distributions, as the majority of the cases investigated prior to the distribution shift are still a part of the new system.

For most technologies the scenario result is relatively close to the mean of the distribution resulting from stochastic optimization, but for some it gives a wrong impression of the expected value compared to stochastic optimization. One example is the expected amount of carbon emissions. For energy planning with the goal of carbon emission reduction, these differences between scenario and stochastic optimization suggest that stochastic optimization may play a crucial role in realistically assessing if emission reductions achieved by market mechanisms will suffice to combat climate change. This holds true even if the internalization of externalized costs for carbon dioxide emissions are considered. The effect of Jensen's inequality can be observed, in all three of the investigated emission reduction cases. The scenario approach overestimates the necessary system cost compared to the stochastic investigation. This effect increases with increasing allowed carbon emissions but is still small with a maximum of 3.5 % in the cases investigated herein.

For some of the technologies as well as carbon emissions and the cost of energy, relatively broad distributions result from stochastic optimization. Therefore a variety of different energy systems with partly very different technology implementations are possible as cost optimal future systems, which can not be accounted for by a scenario that consists only of one possible system layout. The use of a clustering algorithm to group the systems resulting from stochastic optimization by the implemented technologies and their capacities, enables revealing groups of different system layouts, which can be communicated more easily than the combination of all parameter distributions. Also these clusters contain the information of the interactions between the technologies and their combined usage. Analyzing the mean of the uncertain input parameters' distribution by cluster yields additional information as to the input parameter combinations leading to certain clusters. The clustered systems' properties regarding system cost and carbon emissions can also be compared and reveal significant differences between the clusters, i.e. clusters with no or very low carbon emissions and systems with high costs.

Factor effect analysis can be used to gain insights into the cause effect dependencies of the optimized systems. A meta model based on factor effect analysis has been used successful for an approximation of the stochastic optimization results. The quality of the meta model for a certain technology is not known without benchmark and stochastic optimization with the same number of optimizations necessary to derive the meta model many times approximates benchmark results more accurately. Therefore meta models as applied herein only deliver additional value if they are used to either predict extreme cases, which are underrepresented in stochastic optimiziations with reduced number of runs, or to allow for quickly estimating cost distributions in light of newly assessed parameter ranges. The latter application necessitates a broader parameter space for the training data compared to the meta models constructed herein, as these cover only the parameter space of the distribution widths relevant to replicate benchmark results.

8.3 Conclusions, Research Demand and Significance for Decision-Making

Stochastic optimization as performed herein is an opportunity to improve the robustness of ESO results as well as information gain and reliability of the model results. To the mind of the author also trust in results can be improved as uncertainty, which is undoubtedly present, is adequately addressed and accounted for. Cluster analysis allows for additional insights and helps to communicate the results and aid decision making by reducing complexity and focusing on the structure of the energy system instead of single technology combinations. Jensen's inequality means that the probable expected system cost of an optimized system will always be lower than that of the corresponding mean scenario. This effect, although to a limited degree, is present in the examples given in this work. The general doubt regarding the meaningfulness of system cost resulting from an optimization model and its interpretation can be considered more important. Using stochastic optimization with fewer model runs can support the adaption of the use of stochastic optimization as this allows retaining many of the advantages of stochastic optimization at significantly reduced computational effort and relatively small information loss regarding quartiles. This is especially important in the light of increasingly complex models.

Future research is required regarding the impact of uncertainty from demand curves regarding shape and scale, as well as renewable generation curves, e.g. resulting from larger rotor diameters in wind power plants leading to a more even power output of wind turbines. Also the effect of the general uncertainty concerning all technologies including those for which herein fixed prices have been assumed and, taking into account possible future developments, efficiencies should be investigated, as this greatly increases the number of uncertain parameters. Even though their distribution widths are small, a greater influence of Jensen's inequality can be expected based on the results presented in chapter 4. Further investigation of the impact of imposed upper limits of certain technologies, as herein wind turbines, should also be investigated, as e.g. a probable overall system cost increase could be assigned to political decisions determining such boundary conditions. Another aspect of interest is the investigation of disruptive developments in the combination with uncertain input parameters, as this would allow assessing the impact of possible but improbable developments in an uncertain future and the necessity to include these into planning. The derivation of meta models for the whole parameter space of the uncertain parameters is of interest. This could allow for an adaption of the model to changing input parameter distributions and therefore make the results adaptable to assumption changes over time and new insights gained from the inclusion of additional stakeholders. The fast calculation of approximated results would also allow for the implementation of an interactive user interface, e.g. on a website, that enables the interested public or researchers to enter their own assumptions regarding future cost developments. This is relevant as there is usually no consensus regarding the assumptions made in published studies and the impact of the differences in input assumptions on the result is difficult to estimate due to complex and nonlinear dependencies.

Chapter 9

Bibliography

[1]   Narasimha D. Rao and Shonali Pachauri. "Energy access and living standards: some observations on recent trends". In: *Environmental Research Letters* 12.2 (2017), p. 025011. ISSN: 0301-4215. DOI: `10.1088/1748-9326/aa5b0d`. URL: `https://iopscience.iop.org/article/10.1088/1748-9326/aa5b0d`.

[2]   David J. C. Mac Kay. *Sustainable Energy - without the hot air*. Cambridge, 2008. URL: `http://www.inference.org.uk/sustainable/book/tex/sewtha.pdf`.

[3]   Bundesministerium für Wirtschaft und Energie. *Zahlen und Fakten Energiedaten: Nationale und Internationale Entwicklung*. Ed. by Bundesministerium für Wirtschaft und Energie. online, 2020. URL: `https://www.bmwi.de/Redaktion/DE/Binaer/Energiedaten/energiedaten-gesamt-xls.xlsx?__blob=publicationFile&v=121`.

[4]   K. Bithas and P. Kalimeris. "Re-estimating the decoupling effect: Is there an actual transition towards a less energy-intensive economy?" In: *Energy* 51 (2013), pp. 78–84. ISSN: 03605442. DOI: `10.1016/j.energy.2012.11.033`. URL: `http://www.sciencedirect.com/science/article/pii/S0360544212008948`.

[5] Astrid Matthey and Björn Bünger. "Methodenkonvention 3.0 zur Ermittlung von Umweltkosten – Kostensätze". In: (). URL: `https://www.umweltbundesamt.de/sites/default/files/medien/1410/publikationen/2019-02-11_methodenkor 3-0_kostensaetze_korr.pdf`.

[6] Philip Eyrikson Tetlock. *Expert political judgment: How good is it? How can we know?* Princeton, N.J: Princeton University Press, 2005. ISBN: 978-0-691-12302-8. URL: `http://site.ebrary.com/lib/alltitles/docDetail.action?docID=10320496`.

[7] L. K. Kirchmayer. *Economic Operation Of Power Systems.* New York: John Wiley & Sons, Inc., 1958. URL: `https://archive.org/stream/in.ernet.dli.2015.147965/2015.147965.Economic-Operation-Of-Power-Systems_djvu.txt`.

[8] AIEE Committee Report. "System-Planning Practices [includes discussion]". In: *Transactions of the American Institute of Electrical Engineers. Part III: Power Apparatus and Systems* 74.3 (1955). ISSN: 0097-2460. DOI: `10.1109/AIEEPAS.1955.4499168`.

[9] W. G. Chandler, P. L. Dandeno, A. F. Glimn, and L. K. Kirchmayer. "Short-Range Economic Operation of a Combined Thermal and Hydroelectric Power System [includes discussion]". In: *Transactions of the American Institute of Electrical Engineers. Part III: Power Apparatus and Systems* 72.5 (1953), pp. 1057–1065. ISSN: 2379-6766. DOI: `10.1109/AIEEPAS.1953.4498741`.

[10] A. F. Glimn and L. K. Kirchmayer. "Economic Operation of Variable-Head Hydroelectric Plants". In: *Transactions of the American Institute of Electrical Engineers. Part III: Power Apparatus and Systems* 77.3 (1958), pp. 1070–1078. ISSN: 0097-2460. DOI: `10.1109/AIEEPAS.1958.4500100`.

[11] J. H. Drake, L. K. Kirchmayer, R. B. Mayall, and H. Wood. "Optimum Operation of a Hydrothermal System". In: *Transactions of the American Institute of Electrical Engineers. Part III: Power Apparatus and Systems* 81.3 (1962), pp. 242–248. ISSN: 0097-2460. DOI: 10.1109/AIEEPAS.1962.4501309.

[12] George Bernard Dantzig. *Linear Programming and Extensions: R-366-PR.* Santa Monica, California. URL: https://www.rand.org/content/dam/rand/pubs/reports/2007/R366part1.pdf.

[13] Dominique Finon. "Optimisation model for the French energy sector". In: *Energy Policy* 2.2 (1974), pp. 136–151. DOI: 10.1016/0301-4215(74)90005-6.

[14] L. G. Fishbone et al. *MARKAL: A Multiperiod, Linear-Programming Model for Energy Systems Analysis (BNL Version).* Ed. by Brookhaven National Laboratory. Dublin, Ireland.

[15] Leslie G. Fishbone and Harold Abilock. "Markal, a linear-programming model for energy systems analysis: Technical description of the bnl version". In: *International Journal of Energy Research* 5.4 (1981), pp. 353–375. ISSN: 0363907X. DOI: 10.1002/er.4440050406.

[16] Dominik Möst, Wolf Fichtner, and Armin Grunwald, eds. *Energiesystemanalyse: Tagungsband des Workshops "Energiesystemanalyse" vom 27. November 2008 am KIT Zentrum Energie, Karlsruhe.* Karlsruhe: Univ.-verl., 2009. ISBN: 9783866443891.

[17] Manish Ram, Dimitri Bogdanov, Arman Aghahosseini, Solomon Oyewo, Ashish Gulagi, Michael Child, and Christian Breyer. *Global Energy System Based on 100% Renewable Energy - Power Sector: Study by Lappeenranta University of Technology and Energy watch Group.* Berlin. URL: http://energywatchgroup.org/wp-content/uploads/2017/11/Full-Study-100-Renewable-Energy-Worldwide-Power-Sector.pdf.

[18]   H. Turton, V. Panos, M. Densing, and K. Volkart. *Global Multi-regional MARKAL (GMM) model Global Multi-regional MARKAL (GMM) model update: Disaggregation to 15 regions and 2010 recalibration.* Ed. by Paul Scherrer Institut. Villingen, Switzerland. URL: `https://www.psi.ch/sites/default/files/import/eem/PublicationsTa PSI-Bericht_13-03.pdf`.

[19]   R. Loulou, U. Remne, A. Kanudia, A. Lehtila, and G. Goldstein. *Documentation of the TIMES Model: Part I: TIMES Concepts and Theory.* URL: `http://iea-etsap.org/docs/TIMESDoc-Intro.pdf`.

[20]   M. Zonzooz, Z. M. Nopiah, A. M. Yusof, and K. Sopian. "A Review of MARKAL Energy Modeling". In: *European Journal of Scientific Research* 26.3 (2009), pp. 352–361.

[21]   Janusz Cofala. "Modeling of a Medium-Term Development of Country´s Energy System". In: *Annual Review in Automatic Programming* 12.2 (1985), pp. 395–398. URL: `https://www.sciencedirect.com/science/article/abs/pii/0066413885904094?via%3Dihub`.

[22]   Damyant Luthra and J.David Fuller. "Exploring regional energy futures in Canada: A techno-economic energy model for Ontario". In: *Energy* 15.10 (1990), pp. 885–898. ISSN: 03605442. DOI: `10.1016/0360-5442(90)90070-I`.

[23]   Leo Schrattenholzer. "The Energy Supply Model Message". In: *Research Reports IIASA* 1981 (December 1981).

[24]   J.David Fuller and Damyant Luthra. "Model formulation with the Waterloo Energy Modelling System (WATEMS)". In: *Energy* 15.5 (1990), pp. 413–425. ISSN: 03605442. DOI: `10.1016/0360-5442(90)90038-4`. URL: `https://www.sciencedirect.com/science/article/abs/pii/0360544290900384`.

[25] NICLAS MATTSSON and CLAS-OTTO WENE. "Assessing new Energy Technologies Using an Energy System Model with Endogenized Experience Curves". In: *International Journal of Energy Research* 21.4 (1997), pp. 385–393. ISSN: 0363907X. DOI: `10.1002/(SICI)1099-114X(19970325)21:4{\textless}385::AID-ER275{\textgreater}3.0.CO;2-1`.

[26] Bruce Henderson. *The Experience Curve: Reviewed (Part II)*. 1973. URL: `https://www.bcg.com/publications/1973/corporate-finance-strategy-portfolio-management-experience-curve-reviewed-part-ii-the-history.aspx`.

[27] Bo Andersson and Erik Hådén. "Power production and the price of electricity: an analysis of a phase-out of Swedish nuclear power". In: *Energy Policy* 25.13 (1997), pp. 1051–1064. DOI: `10.1016/S0301-4215(97)00085-2`.

[28] Leonardo Barreto. "Technological Learning in Energy Optimisation Models and Deplo". Disseration. Zurich: Siss Federal Institute of Technology Zurich. URL: `https://iea-etsap.org/docs/BARRETO-thesis.pdf`.

[29] Peter Rafaj, Leonardo Barreto, and Socrates Kypreos. "Combining policy instruments for sustainable energy systems: An assessment with the GMM model". In: *Environmental Modeling & Assessment* 11.4 (2006), pp. 277–295. ISSN: 1420-2026. DOI: `10.1007/s10666-005-9037-z`.

[30] Christine Shearer, John Bistline, Mason Inman, and Steven J. Davis. "The effect of natural gas supply on US renewable energy and CO_2 emissions". In: *Environmental Research Letters* 9.9 (2014), p. 094008. ISSN: 0301-4215. DOI: `10.1088/1748-9326/9/9/094008`.

[31] Akram Fakhri Sandvall, Erik O. Ahlgren, and Tomas Ekvall. "Cost-efficiency of urban heating strategies – Modelling scale effects of low-

energy building heat supply". In: *Energy Strategy Reviews* 18 (2017), pp. 212–223. DOI: 10.1016/j.esr.2017.10.003.

[32] Akram Fakhri Sandvall, Erik O. Ahlgren, and Tomas Ekvall. "Low-energy buildings heat supply–Modelling of energy systems and carbon emissions impacts". In: *Environmental Research Letters* 111 (2017), pp. 371–382. ISSN: 0301-4215. DOI: 10.1016/j.enpol.2017.09.007.

[33] O. Bahn and A. Haurie and D.S. Zachary. "Mathematical Modeling And Simulation Methods In Energy Systems". In: (). URL: http://www.eolss.net/sample-chapters/c02/e6-03b-03-06.pdf.

[34] N.M.J.P. van Beeck. "Classification of Energy Models". In: (1999). URL: https://research.tilburguniversity.edu/en/publications/classification-of-energy-models.

[35] Johannes Dorfner et al. *tum-ens/urbs: urbs v1.0.1.* 2019. DOI: 10.5281/ZENODO.3265960.

[36] Benedikt Schweiger. *Cross-sektorale Campus-Konzepte für die Wärmewende: Dokumentation des 6. Projektleitertreffens, Potsdam, 06.-07. Mai 2019.* Schriftenreihe Energiewendebauen, Wissenschaftliche Begleitforschung. Aachen: Wissenschaftliche Begleitforschung Energiewendebauen, RWTH Aachen University, Lehrstuhl für Gebäude- und Raumklimatechnik, 2019. ISBN: 978-3948234-86-7.

[37] J. C. Hourcade, R. Richels, J. R. Robinson, and L. Schrattenholzer. *Estimating the costs of mitigating greenhouse gases.* Climate Change 1995: Economic and Social Dimensions of Climate Change. Cambridge: Cambridge University Press, 1996. ISBN: 0-521-56431-X. URL: http://pure.iiasa.ac.at/id/eprint/4747/.

[38] Michael Grubb, Jae Edmonds, Patrick ten Brink, and Michael Morrison. "The Costs of Limiting Fossil-Fuel CO2 Emissions: A Survey and Analysis". In: *Annual Review of Energy and the Environment* 18.1 (1993), pp. 397–478. ISSN: 1056-3466. DOI: 10.1146/annurev.eg.18.110193.002145.

[39] René Codoni, Hi-chun Park, and K. V. Ramani, eds. *Integrated energy planning: A manual.* Kuala Lumpur, 1985. ISBN: 9679995453.

[40] S. Messner and M. Strubegger. *Model-Based Decision Support in Energy Planning.* Laxenburg, Austria. URL: http://pure.iiasa.ac.at/id/eprint/4477/.

[41] Maxime Kleinpeter. *Energy planning and policy.* Repr. UNESCO energy engineering series Energy engineering learning package. Chichester: Wiley, 1996. ISBN: 0-471-95536-1.

[42] World Energy Conference. *Energy Terminology: A Multi-Lingual Glossary.* 2nd ed. Burlington: Elsevier Science, 1986. ISBN: 978-0080340715. URL: http://gbv.eblib.com/patron/FullRecord.aspx?p=1675205.

[43] M. Zonooz, Z. M. Nopiah, A. M. Yusof, and K. Sopian. "A review of MARKAL energy modeling". In: *European Journal of Scientific Research* 26.3 (2009), pp. 352–361. URL: https://www.researchgate.net/publication/265152928.

[44] L. A. Zadeh. "The concept of a linguistic variable and its application to approximate reasoning—I". In: *Information Sciences* 8.3 (1975), pp. 199–249. ISSN: 00200255. DOI: 10.1016/0020-0255(75)90036-5.

[45] J.G Dijkman, H. van Haeringen, and S.J de Lange. "Fuzzy numbers". In: *Journal of Mathematical Analysis and Applications* 92.2 (1983), pp. 301–341. ISSN: 0022247X. DOI: 10.1016/0022-247X(83)90253-6.

[46] Ali Ebrahimnejad and Madjid Tavana. "A novel method for solving linear programming problems with symmetric trapezoidal fuzzy numbers". In: *Applied Mathematical Modelling* 38.17-18 (2014), pp. 4388–4395. ISSN: 0307904X. DOI: 10.1016/j.apm.2014.02.024.

[47] Reza Ghanbari, Khatere Ghorbani-Moghadam, Nezam Mahdavi-Amiri, and Bernard de Baets. "Fuzzy linear programming problems: models and solutions". In: *Soft Computing* 24.13 (2020), pp. 10043–10073. ISSN: 1433-7479. DOI: 10.1007/s00500-019-04519-w.

[48] Jürgen Branke. *Multiobjective optimization: Interactive and evolutionary approaches.* Vol. 5252. Lecture Notes in Computer Science. Berlin: Springer, 2008. ISBN: 9783540889083. DOI: 10.1007/978-3-540-88908-3. URL: http://site.ebrary.com/lib/alltitles/docDetail.action?docID=10253601.

[49] Babooshka Shavazipour and Theodor J. Stewart. "Multi-objective optimisation under deep uncertainty". In: *Operational Research* (2019). ISSN: 1109-2858. DOI: 10.1007/s12351-019-00512-1.

[50] John R. Birge and François Louveaux. *Introduction to Stochastic Programming.* New York, NY: Springer New York, 2011. ISBN: 978-1-4614-0236-7. DOI: 10.1007/978-1-4614-0237-4.

[51] Khizir Mahmud and Graham E. Town. "A review of computer tools for modeling electric vehicle energy requirements and their impact on power distribution networks". In: *Applied Energy* 172 (2016), pp. 337–359. ISSN: 03062619. DOI: 10.1016/j.apenergy.2016.03.100. URL: http://www.sciencedirect.com/science/article/pii/S0306261916304275.

[52] Millett Granger Morgan and Max Henrion. *Uncertainty: A guide to dealing with uncertainty in quantitative risk and policy analysis.* 1. paperback ed. Cambridge: Cambridge Univ. Press, 1992. ISBN: 0-521-42744-4.

[53] Joseph DeCarolis et al. "Formalizing best practice for energy system optimization modelling". In: *Applied Energy* 194 (2017), pp. 184–198. ISSN: 03062619. DOI: 10.1016/j.apenergy.2017.03.001.

[54] Robert J. Lempert, Steven W. Popper, and Steven C. Bankes. *Shaping the next one hundred years: New methods for quantitative, long-term policy analysis and bibliography*. Santa Monica, CA: RAND, 2003. ISBN: 0833032755. DOI: 10.7249/mr1626rpc. URL: https://www.rand.org/content/dam/rand/pubs/monograph_reports/2007/MR1626.pdf.

[55] Patrick Wrobel, Matthias Schnier, Cornelius Schill, Annedore Kangießer, and Carsten Beier, eds. *Planungshilfsmittel: Praxiserfahrungen aus der energetischen Quartiersplanung*. Stuttgart: Fraunhofer IRB Verlag, 2016. ISBN: 9783816796404. URL: https://www.irbnet.de/daten/rswb/16049026058.pdf.

[56] James V. Beck and Kenneth J. Arnold. *Parameter estimation in engineering and science*. Wiley series in probability and mathematical statistics. New York: Wiley, 1977. ISBN: 0471061182.

[57] Donella H. Meadows. *The limits to growth: A report for the Club of Rome's Project on the Predicament of Mankind*. 2. ed. New York: Universe Books, 1974. ISBN: 0-87663-165-0.

[58] E. G. Nourse, T. N. Carver, G. F. Warren, Charles J. Brand, B. H. Hibbard, and O. C. Stine. "The Outlook for Agriculture". In: *American Journal of Agricultural Economics* 9.1 (1927), pp. 21–52. ISSN: 0002-9092. DOI: 10.2307/1230563. URL: https://academic.oup.com/ajae/article-pdf/9/1/21/187294/9-1-21.pdf.

[59] Erich Jantsch. *Technological Forecasting in Perspective: A framework for technological forecasting, its techniques and organisation: a description of activities and annotated Bibliography by Erich Jantsch*. 1967.

[60]   T. P. Wright. "Factors Affecting the Cost of Airplanes". In: *Journal of the Aeronautical Sciences* 3.4 (1936), pp. 122–128. DOI: `10.2514/8.155`.

[61]   N. Gerdsri. "An Analytical Approach to Building a Technology Development Envelope (TDE) for Roadmapping of Emerging Technologies". In: *International Journal of Innovation and Technology Management* 4.2 (2007), pp. 121–135.

[62]   O. Schmidt, A. Hawkes, A. Gambhir, and I. Staffell. "The future cost of electrical energy storage based on experience rates". In: *Nature Energy* 2.8 (2017), p. 17110. ISSN: 2058-7546. DOI: `10.1038/nenergy.2017.110`. URL: `https://www.nature.com/articles/nenergy2017110.pdf`.

[63]   Fraunhofer ISE. "Current and Future Cost of Photovoltaics. Long-term Scenarios for Market Development, System Prices and LCOE of Utility-Scale PV Systems. Study on behalf of Agora Energiewende". In: (2015).

[64]   Lena Neij. "Cost development of future technologies for power generation—A study based on experience curves and complementary bottom-up assessments". In: *Energy Policy* 36.6 (2008), pp. 2200–2211. DOI: `10.1016/j.enpol.2008.02.029`. URL: `http://www.sciencedirect.com/science/article/pii/S030142150`

[65]   Aüirba Sakti, Ines M.L. Azevedo, Erica R.H. Fuchs, Jeremy J. Michalek, Kevin G. Gallagher, and Jay F.. Whitacre. "Consistency and robustness of forecasting for emerging technologies: The case of Li-ion batteries for electric vehicles". In: *Energy Policy* 106 (2017), pp. 415–426. URL: `Science%20Direct`.

[66]   Priscila Gonçalves Vasconcelos Sampaio, Mário Orestes Aguirre González, Rafael Monteiro de Vasconcelos, Marllen Aylla Teixeira dos Santos, Priscila da Cunha Jácome Vidal, Jonathan Paulo Pinheiro Pereira, and Everton Santi. "Prospecting technologies for photovoltaic solar energy: Overview of its technical–commercial viability". In: *International*

Journal of Energy Research 44.2 (2020), pp. 651–668. ISSN: 0363907X. DOI: 10.1002/er.4957.

[67] Norman Crolee Dalkey and Olaf Helmer-Hirschberg. *An Experimental Application of the Delphi Method to the Use of Experts.* Ed. by The RAND Corporation. URL: https://www.rand.org/content/dam/rand/pubs/research_memoranda/2009/RM727.1.pdf.

[68] Fabiana Scapolo and Ian Miles. "Eliciting experts' knowledge: A comparison of two methods". In: *Technological Forecasting and Social Change* 73.6 (2006), pp. 679–704. ISSN: 00401625. DOI: 10.1016/j.techfore.2006.03.001.

[69] Joseph Voros. "A generic foresight process framework". In: *Foresight* 5.3 (2003), pp. 10–21. ISSN: 1463-6689. DOI: 10.1108/14636680310698379.

[70] BMBF-Internetredaktion. *Foresight als Methode der Strategischen Vorausschau im BMBF - BMBF.* 2020. URL: https://www.bmbf.de/de/die-philosophie-dahinter-936.html.

[71] *PRECURSOR | Bedeutung im Cambridge Englisch Wörterbuch.* 20.07.2020. URL: https://dictionary.cambridge.org/de/worterbuch/englisch/precursor.

[72] G. M. Ballard. "Precursor Analysis—a way of testing safety analysis?" In: *Reliability Engineering & System Safety* 27.1 (1990), pp. 77–89. ISSN: 09518320. DOI: 10.1016/0951-8320(90)90032-I.

[73] Office of Saftey and Mission Assurance. *Accident Precursor Analysis Handbook: Version 1.0.* Ed. by National Aeronautics and Space Administration. Washington, D.C., 2011. URL: https://ntrs.nasa.gov/archive/nasa/casi.ntrs.nasa.gov/20120003292.pdf.

[74] Thomas Rose and Trevor Sweeting. "How safe is nuclear power? A statistical study suggests less than expected". In: *Bulletin of the Atomic Scientists* 72.2 (2016), pp. 112–115. ISSN: 0096-3402. DOI: 10.1080/00963402.2016.1145910.

[75] Walk and R. Steven. "Quantitative Technology Forecasting Techniques". In: *Technological Change*. Ed. by Aurora Teixeira. InTech, 2012. ISBN: 978-953-51-0509-1. DOI: 10.5772/38024.

[76] Eirini Ozouni, Constantinos Katrakylidis, and Grigoris Zarotiadis. "Technology evolution and long waves: investigating their relation with spectral and cross-spectral analysis". In: *Journal of Applied Economics* 21.1 (2018), pp. 160–174. ISSN: 1514-0326. DOI: 10.1080/15140326.2018.1526872.

[77] Sofi Kurki. "The Long-Waves and the Evolution of Futures Practice and Theory". In: *World Futures Review* 11.2 (2019), pp. 122–140. ISSN: 1946-7567. DOI: 10.1177/1946756718796487.

[78] Dmitry Kucharavy and Roland DeGuio. "Application of S-Shaped Curves". In: *TRIZ-Future Conference 2007: Current Scientific and Industrial Reality* (2008). URL: https://hal.archives-ouvertes.fr/file/index/docid/282758/filename/TCF2007_DK_RDG.pdf.

[79] Mikael Höök, Junchen Li, Noriaki Oba, and Simon Snowden. "Descriptive and Predictive Growth Curves in Energy System Analysis". In: *Natural Resources Research* 20.2 (2011), pp. 103–116. ISSN: 1520-7439. DOI: 10.1007/s11053-011-9139-z.

[80] Mohamad Y. Jaber. *Learning curves: Theory, models, and applications.* Vol. 21. Industrial innovation series. Boca Raton, FL: CRC Press, 2011. ISBN: 978-1-4398-0740-8. URL: http://site.ebrary.com/lib/academiccompletetitles/home.action.

[81] Sascha Samadi. "The experience curve theory and its application in the field of electricity generation technologies – A literature review". In: *Renewable and Sustainable Energy Reviews* 82 (2018), pp. 2346–2364. ISSN: 13640321. DOI: 10.1016/j.rser.2017.08.077. URL: https://epub.wupperinst.org/frontdoor/deliver/index/docId/6806/file/6806_Samadi.pdf.

[82] Gordon and Theodore Jay. "Trend impact analysis". In: *Futures Research Methodology* (1994). URL: http://www.foresight.pl/assets/downloads/publications/Gordon1994-Trendimpact.pdf.

[83] Kelsey A. W. Horowitz, Michael Woodhouse, Hohyun Lee, and Greg P. Smestad. "A bottom-up cost analysis of a high concentration PV module". In: AIP Conference Proceedings. AIP Publishing LLC, 2015, p. 100001. DOI: 10.1063/1.4931548. URL: https://aip.scitation.org/doi/pdf/10.1063/1.4931548.

[84] Lena Neij, Mads Borup, Markus Blesl, and Oliver Mayer-Spohn. *Cost development - an analysis based on experience curves: Deliverable D 3 3 - RS 1a: NEEDS - New Energy Externalities Developments for Sustainability.* 2016. URL: http://www.needs-project.org/RS1a/Deliverable%20D%203%203%20-%20RS%201a%20(3).pdf.

[85] Christopher Cherniak, Richard Nisbett, and Lee Ross. "Human Inference: Strategies and Shortcomings of Social Judgment". In: *The Philosophical Review* 92.3 (1983), p. 462. ISSN: 00318108. DOI: 10.2307/2184495.

[86] Amos Tversky and Daniel Kahneman. "Availability: A heuristic for judging frequency and probability". In: *Cognitive Psychology* 5.2 (1973), pp. 207–232. ISSN: 0010-0285. DOI: 10.1016/0010-0285(73)

90033 - 9. URL: `http://www.sciencedirect.com/science/article/pii/0010028573900339`.

[87] Reid Hastie and Robyn M. Dawes. *Rational choice in an uncertain world: The psychology of judgment and decision making*. [7. Nachdr.] Thousand Oaks, Calif.: Sage Publ, 2001. ISBN: 0-7619-2275-X.

[88] Amos Tversky and Daniel Kahneman. "Belief in the law of small numbers". In: *Psychological Bulletin* 76.2 (1971), pp. 105–110. ISSN: 0033-2909. DOI: `10.1037/h0031322`.

[89] Daniel Kahneman and Amos Tversky. "Subjective probability: A judgment of representativeness". In: *Cognitive Psychology* 3.3 (1972), pp. 430–454. ISSN: 0010-0285. DOI: `10.1016/0010-0285(72)90016-3`. URL: `http://www.sciencedirect.com/science/article/pii/0010028572900163`.

[90] Daniel Kahneman and Amos Tversky. "On the psychology of prediction". In: *Psychological Review* 80.4 (1973), pp. 237–251. ISSN: 0033-295X. DOI: `10.1037/h0034747`.

[91] Philip Eyrikson Tetlock and Dan Gardner. *Superforecasting: The art and science of prediction*. First edition. New York: Crown Publishers, 2015. ISBN: 0804136696.

[92] Barbara Mellers et al. "Identifying and cultivating superforecasters as a method of improving probabilistic predictions". In: *Perspectives on psychological science : a journal of the Association for Psychological Science* 10.3 (2015), pp. 267–281. DOI: `10.1177/1745691615577794`.

[93] *Stochastic Optimization – from Wolfram MathWorld*. 27.07.2020.

[94] M. G. Morgan, H. Dowlatabadi, Henrion M., D. Keith, R. Lempert, S. McBride, M. Small, and T. Wilbanks. *Best Practice Approaches for Characterizing, Communicating and Incorporating Scientific Uncertainty in Climate Decision Making*. Ed. by National Oceanic and Atmospheric Administration.

[95] Wolf Wedel, Andreas Hanel, Harmut Spliehoff, and Annelies Vandersickel. "Improving information gain from optimization problems using artificial neural networks". In: *Proceedings of ECOS 2019* (2019).

[96] Wolf Wedel. "A false sense of certainty: Optimization without considering input parameter distributions - the case of energy systems". In: *Manuscript submitted for publication* (2020).

[97] Steve R. Carpenter. *Ecosystems and human well-being*. Millennium ecosystem assessment series. Washington, DC: Island Press, 2005. ISBN: 9781559633901.

[98] Lauge Baungaard Rasmussen. "The narrative aspect of scenario building - How story telling may give people a memory of the future". In: *AI & SOCIETY* 19.3 (2005), pp. 229–249. ISSN: 0951-5666. DOI: `10.1007/s00146-005-0337-2`. URL: `https://link.springer.com/content/pdf/10.1007/s00146-005-0337-2.pdf`.

[99] Mark D. A. Rounsevell and Marc J. Metzger. "Developing qualitative scenario storylines for environmental change assessment". In: *Wiley Interdisciplinary Reviews: Climate Change* 1.4 (2010), pp. 606–619. ISSN: 17577780. DOI: `10.1002/wcc.63`.

[100] Nick Hughes. *Transition Pathways to a low Carbon Economy: A Historical Overview of Strategic Scenario Planning: A Joint Working Paper of the UKERC and the EON.UK/EPSRC Transition Pathways Project.* 2009. URL: `https://www.researchgate.net/publication/238111229_A_Historical_Overview_of_Strategic_Scenario_Planning_A_Joint_Working_Paper_of_the_UKERC_and_the_EONUKEPSRC_Transition_Pathways_Project.`

[101] Andrea Saltelli. "Sensitivity analysis: Could better methods be used?" In: *Journal of Geophysical Research: Atmospheres* 104.D3 (1999), pp. 3789–3793. ISSN: 01480227. DOI: `10.1029/1998JD100042`.

[102] Francesca Campolongo, Jessica Cariboni, and Andrea Saltelli. "An effective screening design for sensitivity analysis of large models". In: *Environmental Modelling & Software* 22.10 (2007), pp. 1509–1518. ISSN: 13648152. DOI: 10.1016/j.envsoft.2006.10.004.

[103] Karl Siebertz, David van Bebber, and Thomas Hochkirchen. *Statistische Versuchsplanung*. Berlin, Heidelberg: Springer Berlin Heidelberg, 2010. ISBN: 978-3-642-05492-1. DOI: 10.1007/978-3-642-05493-8.

[104] R. Ciampalini, S. Follain, B. Cheviron, Y. Le Bissonnais, A. Couturier, R. Moussa, and C. Walter. "Local Sensitivity Analysis of the Land-Soil Erosion Model Applied to a Virtual Catchment". In: *Sensitivity Analysis in Earth Observation Modelling*. Elsevier, 2017, pp. 55–73. ISBN: 9780128030110. DOI: 10.1016/B978-0-12-803011-0.00003-3.

[105] Bram L. Gorissen, İhsan Yanıkoğlu, and Dick den Hertog. "A practical guide to robust optimization". In: *Omega* 53 (2015), pp. 124–137. ISSN: 03050483. DOI: 10.1016/j.omega.2014.12.006.

[106] Beate Rhein. "Robuste Optimierung mit Quantilmaßen auf globalen Metamodellen". Disseration. Köln: Universität zu Köln, 15.01.2014. URL: https://kups.ub.uni-koeln.de/5604/1/Diss_Rhein_final.pdf.

[107] A. L. Soyster. "Technical Note—Convex Programming with Set-Inclusive Constraints and Applications to Inexact Linear Programming". In: *Operations Research* 21.5 (1973), pp. 1154–1157. ISSN: 0030-364X. DOI: 10.1287/opre.21.5.1154.

[108] Dimitris Bertsimas, Eugene Litvinov, Xu Andy Sun, Jinye Zhao, and Tongxin Zheng. "Adaptive Robust Optimization for the Security Constrained Unit Commitment Problem". In: *IEEE Transactions on Power Systems* 28.1 (2013), pp. 52–63. ISSN: 1558-0679. DOI: 10.1109/TPWRS.2012.2205021.

[109]   Stefano Moret, Michel Bierlaire, and François Maréchal. "Robust Optimization for Strategic Energy Planning". In: *Informatica* 27.3 (2016), pp. 625–648. DOI: 10.15388/Informatica.2016.103.

[110]   Albert Madansky. "Inequalities for Stochastic Linear Programming Problems". In: *Management Science* 6.2 (1960), pp. 197–204. ISSN: 0025-1909. DOI: 10.1287/mnsc.6.2.197.

[111]   Benjamin F. Hobbs. "Optimization methods for electric utility resource planning". In: *European Journal of Operational Research* 83.1 (1995), pp. 1–20. ISSN: 03772217. DOI: 10.1016/0377-2217(94)00190-N.

[112]   John E. Bistline. "Electric sector capacity planning under uncertainty: Climate policy and natural gas in the US". In: *Energy Economics* 51 (2015), pp. 236–251. ISSN: 0140-9883. DOI: 10.1016/j.eneco.2015.07.008. URL: https://www.sciencedirect.com/science/article/pii/S0140988315002157.

[113]   Christer O. Kiselman and Shiva Samieinia. "Convexity of marginal functions in the discrete case". In: *Analysis Meets Geometry*. Ed. by Mats Andersson, Jan Boman, Christer Kiselman, Pavel Kurasov, and Ragnar Sigurdsson. Trends in Mathematics. Cham: Birkhäuser, 2017, pp. 287–309. ISBN: 978-3-319-52469-6. DOI: 10.1007/978-3-319-52471-9{\textunderscore}18.

[114]   Harlan D. Mills. "Marginal Values of Matrix Games and Linear Programs". In: *Linear Inequalities and Related Systems*. (AM-38) (1956), pp. 183–194. URL: www.jstor.org/stable/j.ctt1b9x27g.11.

[115]   Akshay Agrawal, Robin Verschueren, Steven Diamond, and Stephen Boyd. "A rewriting system for convex optimization problems". In: *Journal of Control and Decision* 5.1 (2018), pp. 42–60. ISSN: 2330-7706. DOI: 10.1080/23307706.2017.1397554.

[116]   Stephen P. Boyd and Lieven Vandenberghe. *Convex Optimization*. 7. printing with corrections. Cambridge: Cambridge Univ. Press, 2009. ISBN: 9780521833783.

[117]   J. L. W. V. Jensen. "Sur les Fonctions Convexes et les Inégalités Entre les Valeurs Moyennes". In: *Acta Mathematica* 30.0 (1906), pp. 175–193. ISSN: 0001-5962. DOI: 10.1007/BF02418571.

[118]   Tristan Needham. "A Visual Explanation of Jensen's Inequality". In: *The American Mathematical Monthly* 100.8 (1993), p. 768. ISSN: 00029890. DOI: 10.2307/2324783.

[119]   Simon Benninga and Aris Protopapadakis. "Real and Nominal Interest Rates under Uncertainty: The Fisher Theorem and the Term Structure". In: *Journal of Political Economy* 91.5 (1983), pp. 856–867. ISSN: 0022-3808. DOI: 10.1086/261185.

[120]   *The monetary policy of the ECB*. 3. ed. Eurosystem. Frankfurt am Main: European Central Bank, 2011. ISBN: 9289907789. URL: https://www.ecb.europa.eu/pub/pdf/other/monetarypolicy2(pdf?4004e7099b3dcdbf58d0874f6eab650e.

[121]   *Rendite zehnjähriger Staatsanleihen Deutschlands - Monatswerte 2020 | Statista: Mai 2019 - Mai 2020, based on data from Bloomberg*. 2020. URL: https://de.statista.com/statistik/daten/studie/238018/umfrage/rendite-zehnjaehriger-staatsanleihen-in-deutschland-nach-monaten/.

[122]   *Datenmeldungen 1. August 2014 bis 30. Juni 2017*. 2019. URL: https://www.bundesnetzagentur.de/DE/Sachgebiete/ElektrizitaetundGas/Unternehmen_Institutionen/ErneuerbareEnergien/ZahlenDatenInformationen/EEG_Registerdaten/ArchivDatenMeldgn/ArchivDatenN node.html.

[123] AGEB Energiebilanzen e.V. *Energieverbrauch: in Deutschland im Jahr 2019*. Ed. by Arbeitsgemeinschaft Energiebilanzen.

[124] Klaus Fichter and Ralph Hintemann. "Beyond Energy". In: *Journal of Industrial Ecology* 18.6 (2014), pp. 846–858. ISSN: 10881980. DOI: `10.1111/jiec.12155`. URL: `https://www.borderstep.de/wp-content/uploads/2017/03/Borderstep_Rechenzentren_2016.pdf`.

[125] Wissenschaftliche Dienste des Deutschen Bundestages. *Energieverbrauch von Rechenzentren: Sachstand.* Ed. by Wissenschaftliche Dienste des Deutschen Bundestages. URL: `https://www.bundestag.de/resource/blob/651446/d226ff9ff67a3c29d893859121cfc WD-8-041-19-pdf-data.pdf`.

[126] Bundesministerium für Wirtschaft und Energie. *Mehr aus Energie machen - Nationaler Aktionsplan für Energieeffizienz.* Ed. by Bundesministerium für Wirtschaft und Energie. online. URL: `https://www.bmwi.de/Redaktion/DE/Publikationen/Energie/nationaler-aktionsplan-energieeffizienz-nape.pdf?__blob=publicationFile&v=10`.

[127] Ya Wu, Qianwen Zhu, and Bangzhu Zhu. "Comparisons of decoupling trends of global economic growth and energy consumption between developed and developing countries". In: *Environmental Research Letters* 116 (2018), pp. 30–38. ISSN: 0301-4215. DOI: `10.1016/j.enpol.2018.01.047`.

[128] Richard York. "Demographic trends and energy consumption in European Union Nations, 1960–2025". In: *Social Science Research* 36.3 (2007), pp. 855–872. ISSN: 0049089X. DOI: `10.1016/j.ssresearch.2006.06.007`.

[129] Maximilian Blume, Andreas Hanel, Sebastian Miehling, Benedikt Schweiger, Jakob Schweiger, René Schwermer, and Wolf Wedel. *100 % erneuerbare Energieversorgung 2040 in Bayern: Beauftragt vom Bund für Umwelt und*

Naturschutz Deutschland e. V. Ed. by Kompetenzzentrum ... Garching, Germany, 2020.

[130]   Bundesanstalt für Straßenwesen. *Fachthemen - Automatische Zählstellen auf Autobahnen und Bundesstraßen.* online, 2020. URL: `https://www.bast.de/DE/Verkehrstechnik/Fachthemen/v2-verkehrszaehlung/Stundenwerte.html;jsessionid0EBD0F74464BC86B20348A3247A449AD.live11294?nn=626916.`.

[131]   Mathias Brandt. "Neuzulassungsrekord bei SUVs". In: *Statista* (8.10.2019). URL: `https://de.statista.com/infografik/19572/anzahl-der-neuzulassungen-von-suv-in-deutschlan`

[132]   Severin Hänggi, Philipp Elbert, Thomas Bütler, Urs Cabalzar, Sinan Teske, Christian Bach, and Christopher Onder. "A review of synthetic fuels for passenger vehicles". In: *Energy Reports* 5 (2019), pp. 555–569. ISSN: 23524847. DOI: `10.1016/j.egyr.2019.04.007`.

[133]   Kraftfahrtbundesamt. *Elektroautos - Anzahl der Neuzulassungen bis 2020 [Graph]: In Statista.* Statista, April 2020. URL: `https://de.statista.com/statistik/daten/studie/244000/umfrage/neuzulassungen-von-elektroautos-in-deutschland/#statisticContainer`.

[134]   Deepak Ronanki, Apoorva Kelkar, and Sheldon S. Williamson. "Extreme Fast Charging Technology—Prospects to Enhance Sustainable Electric Transportation". In: *Energies* 12.19 (2019), p. 3721. ISSN: 1996-1073. DOI: `10.3390/en12193721`.

[135]   *200 kilometers in 8 minutes: ABB's fast chargers power the e-mobility revolution.* 20.04.2020. URL: `https://new.abb.com/news/detail/4996/ABBs-fast-chargers-power-the-e-mobility-revolution`.

[136] *EV Supercharging - Petalite.* 20.04.2020. URL: `https://www.petalite.co.uk/evsupercharging/`.

[137] Umweltbundesamt (UBA), Fachgebiet V 1.4, und BMWi. *Energieeffizienz in Zahlen 2019 - Entwicklungen und Trends in Deutschland 2019.* Ed. by Bundesministerium für Wirtschaft und Energie. online. URL: `https://www.bmwi.de/Redaktion/DE/Publikationen/Energie/energieeffizienz-in-zahlen-2019.pdf%3F__blob%3DpublicationFile%26v%3D8`.

[138] Umwelt Bundesamt. *Energiebedingte Emissionen.* Ed. by Umwelt Bundesamt. online, 2020. URL: `https://www.umweltbundesamt.de/daten/energie/energiebedingte-emissionen#energiebedingte-treibhausgas-emissionen`.

[139] Sabine Radke. *Verkehr in Zahlen 2019/2020.* Ed. by Bundesministerium für Verkehr und digitale Infrastruktur. Flensburg. URL: `https://www.bmvi.de/SharedDocs/DE/Publikationen/G/verkehr-in-zahlen-2019-pdf.pdf?__blob=publicationFile`.

[140] Martin Schmied and Moritz Mottschall. *Berechnung des Energieverbrauchs und der Treibhausgasemissionen des ÖPNV: INFRAS – Forschung und Beratung.* Ed. by Bundesministerium für Verkehr und digitale Infrastruktur. URL: `https://www.bmvi.de/SharedDocs/DE/Anlage/G/energieverbrauch-treibhausgasemission-oepnv.pdf?__blob=publicationFile`.

[141] Christoph Falter, Niklas Scharfenberg, and Antoine Habersetzer. "Geographical Potential of Solar Thermochemical Jet Fuel Production". In: *Energies* 13.4 (2020), p. 802. ISSN: 1996-1073. DOI: `10.3390/en13040802`.

[142] Christoph Falter, Valentin Batteiger, and Andreas Sizmann. "Climate Impact and Economic Feasibility of Solar Thermochemical Jet Fuel Production". In: *Environmental science & technology* 50.1 (2016), pp. 470–477. DOI: `10.1021/acs.est.5b03515`.

[143]   Bruce D. Patterson, Frode Mo, Andreas Borgschulte, Magne Hillestad, Fortunat Joos, Trygve Kristiansen, Svein Sunde, and Jeroen A. van Bokhoven. "Renewable CO2 recycling and synthetic fuel production in a marine environment". In: *Proceedings of the National Academy of Sciences of the United States of America* 116.25 (2019), pp. 12212–12219. DOI: 10.1073/pnas.1902335116.

[144]   Andreas W. Schäfer, Steven R. H. Barrett, Khan Doyme, Lynnette M. Dray, Albert R. Gnadt, Rod Self, Aidan O'Sullivan, Athanasios P. Synodinos, and Antonio J. Torija. "Technological, economic and environmental prospects of all-electric aircraft". In: *Nature Energy* 4.2 (2019), pp. 160–166. ISSN: 2058-7546. DOI: 10.1038/s41560-018-0294-x. URL: https://www.nature.com/articles/s41560-018-0294-x.

[145]   Spritmonitor.de. *Spritverbrauch berechnen und Autokosten verwalten.* 2020. URL: https://www.spritmonitor.de/.

[146]   Allgemeiner Deutscher Automobil-Club e.V. *Elektroauto Reichweiten & Stromverbrauch im Vergleich | ADAC.* 2020. URL: https://www.adac.de/rund-ums-fahrzeug/tests/elektromobilitae stromverbrauch-elektroautos-adac-test/.

[147]   Mercedes-Benz. *eVito Kastenwagen | Technische Daten | Mercedes-Benz.* 2020. URL: https://www.mercedes-benz.de/vans/de/vito/e-vito-panel-van/technical-data.

[148]   Mercedes-Benz. *eSprinter Kastenwagen | Technische Daten | Mercedes-Benz.* 2020. URL: https://www.mercedes-benz.de/vans/de/sprinter/e-sprinter-panel-van/technical-data.

[149]   Tesla Inc. *Tesla Semi.* 2019. URL: https://www.tesla.com/semi.

[150]   Volvo Group Trucks Central Europe GmbH. *Volvo FL Electric.* 16.04.2020.
URL: `https://www.volvotrucks.de/de-de/trucks/alternative-antriebe/elektro-lkw/volvo-fl-electric.html`.

[151]   MAN Truck and Bus AG. *Porsche übernimmt eTruck von MAN für umweltfreundliche Logistik | MAN Lkw.* 21.04.2020. URL: `https://www.truck.man.eu/de/de/man-welt/man-in-deutschland/presse-und-medien/Porsche-uebernimmt-eTruck-von-MAN-fuer-umweltfreundliche-Logistik--354432.html`.

[152]   Mercedes-Benz. *Mercedes-Benz Trucks zieht Zwischenbilanz: Elektro-Lkw eActros seit über einem Jahr erfolgreich im Kundeneinsatz - Daimler Global Media Site.* 2019. URL: `https://media.daimler.com/marsMediaSite/de/instance/ko.xhtml?oid=44842543&ls=L2RlL2luc3RhbmNlL2tvLnhodG1sP29pZD0zMzQ1MT!&rs=0`.

[153]   electrive. *DAF rückt Hybrid- und Elektro-Lkw ins Rampenlicht - electrive.net.* 2018. URL: `https://www.electrive.net/2018/09/25/daf-rueckt-hybrid-und-elektro-lkw-ins-rampenlicht/`.

[154]   Kraftfahrt-Bundesamt. *Kraftfahrt-Bundesamt - Fahrzeuge - Bestand an Pkw in den Jahren 2011 bis 2020 nach ausgewählten Kraftstoffarten.* 2019. URL: `https://www.kba.de/DE/Statistik/Fahrzeuge/Bestand/Umwelt/fz_b_umwelt_archiv/2020/2020_b_umwelt_z.html?nn=2601598`.

[155]   Yue Wang, David Infield, and Simon Gill. "Smart charging for electric vehicles to minimise charging cost". In: *Proceedings of the Institution of Mechanical Engineers, Part A: Journal of Power and Energy* 231.6 (2017), pp. 526–534. ISSN: 0957-6509. DOI: `10.1177/0957650916688409`.

[156] Mark Hellwig. "Entwicklung und Anwendung parametrisierer Standard-Lastprofile". Dissertation. München: Technische Universität München, 15.09.2003.

[157] Uwe Bigalke, Aline Armbruster, Franziska Lukas, Oliver Krieger, Cornelia Schuch, and Jan Kunde. *Der dena-Gebäudereport 2016: Statistiken und Analysen zur Energieeffizienz im Gebäudebestand.* Ed. by Deutsche Energie-Agentur GmbH. Berlin.

[158] Blesl et al. *Wärmeatlas Baden-Württemberg: Erstellung eines Leitfadens um Umsetzung für Modellregionen.* 2008.

[159] S. Frisch, M. Pehnt, P. Otter, and M. Nast. *Prozesswärme im Marktanreizprogramm: Zwischenbericht zu Perspektivische Weiterentwicklung des Marktanreizprogramms FKZ 03MAP123.* Heidelberg, Stuttgart, Germany, 2010. URL: `https://elib.dlr.de/82173/1/Prozessw%C3%A4rme_im_MAP.pdf`.

[160] *SMARD - Strommarktdaten, Stromhandel und Stromerzeugung in Deutschland.* 16.04.2020. URL: `https://www.smard.de/home/46`.

[161] Burger, Bruno Prof., Dr. *Installierte Leistung | Energy Charts.* 14.02.2020. URL: `https://www.energy-charts.de/power_inst_de.htm?year=2018&period=monthly&type=power_inst`.

[162] Bundesnetzagentur. *Bundesnetzagentur - Veröffentlichung von EEG-Registerdaten: Veröffentlichung der Registerdaten -08/2014 bis 01/2019.* 22.04.2020. URL: `https://www.bundesnetzagentur.de/DE/Sachgebiete/ElektrizitaetundGas/Unternehmen_Institutionen/ErneuerbareEnergien/ZahlenDatenIn EEG_Registerdaten/EEG_Registerdaten_node.html`.

[163] Andrew Lee. *Offshore wind power price plunges by a third in a year: BNEF | Recharge: Tifenn Brandily for BNEF cited by Recharge.* online, 2019. URL: `https://www.rechargenews.com/transition/`

offshore-wind-power-price-plunges-by-a-third-
in-a-year-bnef/2-1-692944.

[164] Bundesnetzagentur. *WindSeeG - 2. Ausschreibung für bestehende Pro-
jekte nach § 26 WindSeeG: Ergebnisse der 2. Ausschreibung vom 01.04.2018.*
Ed. by Bundesnetzagentur. online. URL: https://www.bundesnetzagentur.
de/DE/Service-Funktionen/Beschlusskammern/
1_GZ/BK6-GZ/2018/BK6-18-001/Ergebnisse_
zweite_ausschreibung.pdf?__blob=publicationFile&
v=3.

[165] IRENA. *Renewable Power Generation Costs in 2017.* Ed. by International
Renewable Energy Agency. Abu Dhabi, 2018.

[166] The International Renewable Energy Agency. *Renewable power generation costs in 2018.* 2019. URL: https://www.irena.org/-
/media/Files/IRENA/Agency/Publication/2019/
May/IRENA_Renewable-Power-Generations-Costs-
in-2018.pdf.

[167] International Renewable Energy Agency. *Renewable power genera-
tion costs in 2019.* 2020. URL: https://irena.org/-/media/
Files/IRENA/Agency/Publication/2020/Jun/IRENA_
Power_Generation_Costs_2019.pdf.

[168] Fraunhofer Institut für Windenergie und Energiesystemtechnik, ed.
Windenergie Report Deutschland 2011. Kassel, 2012. URL: https://
www.fraunhofer.de/content/dam/zv/de/forschungsthemen/
energie/Windreport-2011-de.pdf.

[169] Stefan Bofinger, Doron Callies, Michael Scheibe, Yves-Marie Saint-
Drenan, and Kurt Rohring. *Potential der Windenergienutzung an Land:
Kurzfassung.* Ed. by Bundesverband WindEnergie e.V. Berlin, 2012.
URL: https://www.wind-energie.de/fileadmin/
redaktion/dokumente/publikationen-oeffentlich/

themen/01-mensch-und-umwelt/03-naturschutz/
bwe_potenzialstudie_kurzfassung_2012-03.pdf.

[170]   Hans-Martin Henning and Andreas Palzer. *Studie: 100 % Erneuerbare Energien für Strom und Wärme in Deutschland.*

[171]   Harry Wirth. *Recent Facts about Photovoltaics in Germany.* Ed. by Fraunhofer ISE. URL: https://www.ise.fraunhofer.de/content/dam/ise/en/documents/publications/studies/recent-facts-about-photovoltaics-in-germany.pdf.

[172]   Bundesministerium für Verkehr und digitale Infrastruktur, ed. *Räumlich differenzierte Flächenpotentiale für erneuerbare Energien in Deutschland.* Berlin. URL: https://d-nb.info/1075812623/34.

[173]   Thomas Klaus, Carla Vollmer, Kathrin Werner, Harry Lehmann, and Klaus Müschen. *Energieziel 2050: 100% Strom aus erneuerbaren Quellen.* Ed. by Umwelt Bundesamt. Dessau-Roßlau, 2010. URL: https://www.umweltbundesamt.de/sites/default/files/medien/378/publikationen/energieziel_2050.pdf.

[174]   *Beendete Ausschreibungen: Statistiken - Solar-Anlagen.* 2020. URL: https://www.bundesnetzagentur.de/DE/Sachgebiete/ElektrizitaetundGas/Unternehmen_Institutionen/Ausschreibungen/Solaranlagen/BeendeteAusschreibu BeendeteAusschreibungen_node.html.

[175]   *Beendete Ausschreibungen: Statistiken - Windenergieanlagen an Land.* 2020. URL: https://www.bundesnetzagentur.de/DE/Sachgebiete/ElektrizitaetundGas/Unternehmen_Institutionen/Ausschreibungen/Wind_Onshore/BeendeteAusschreibungen/BeendeteAusschreibungen_node.html.

[176] Sonal Patel. *High-Volume Hydrogen Gas Turbines Take Shape.* 2019. URL: https : / / www . powermag . com / high - volume - hydrogen-gas-turbines-take-shape/.

[177] Paolo Chiesa, Giovanni Lozza, and Luigi Mazzocchi. "Using Hydrogen as Gas Turbine Fuel". In: *Journal of Engineering for Gas Turbines and Power* 127.1 (2005), pp. 73–80. ISSN: 0742-4795. DOI: 10 . 1115 / 1 . 1787513.

[178] Gas Turbine World. *2016-17 GTW Handbook.*

[179] Malta Resources Authority, Lahmeyer International, ed. *Energy Interconnection Europe - Malta: Final Report Work Package 2A.* 2008. DOI: 10 . 14714 / CP59 . 241.

[180] U.S. Energy Information Administration. *Capital Cost Estimates for Utility Scale Electricity Generating Plants.* URL: www . eia . gov.

[181] Acil Allen Consulting. *Fuel and Technology Cost Review: Final Report.* Ed. by Acil Allen Consulting.

[182] PB New Zealand Ltd. *Thermal Power Station Advice - Fixed & Variable O&M Costs: Report for the Electricity Commission.*

[183] ASUE e.V. *BHKW-Kenndaten 2014/2015.* Berlin, 2015.

[184] 2G Energy AG. *Hydrogen CHP - 2G.* 2020. URL: https : //www . 2 - g . com/en/hydrogen-chp/.

[185] 2G Energy AG. *agenitor | 75 to 450 kW - 2G.* 2020. URL: https : // www . 2 - g . com/en/agenitor-75-to-450-kw/.

[186] VDI 2067-30. *Wirtschaftlichkeit gebäudetechnischer Anlagen Energieaufwand der Verteilung.* 6.2013.

[187] Hans Böhm, Andreas Zauner, Sebastian Goers, Robert Tichler, and Pieter Kroon. *Innovative large-scale energy storage technologies and Power-to-Gas concepts after optimization: D7.5 Report on experience curves and economies of scale.* 2018. URL: https://www.storeandgo.info/fileadmin/downloads/deliverables_2019/20190801-STOREandGO-D7.7-EIL-Analysis_on_future_technology_options_and_on_techno-economic_optimization.pdf.

[188] Andreas Zauner, Hans Böhm, Daniel C. Rosenfeld, and Robert Tichler. *Innovative large-scale energy storage technologies and Power-to-Gas concepts after optimization: D7.7 Analysis of future technology options and on techno-economic options.* 2019. URL: https://www.storeandgo.info/fileadmin/downloads/deliverables2019/20190801-STOREandGO-D7.7-EIL-Analysis_on_future_technology_options_and_on_techno-economic_optimization.pdf.

[189] Alexander Buttler and Hartmut Spliethoff. "Current status of water electrolysis for energy storage, grid balancing and sector coupling via power-to-gas and power-to-liquids: A review". In: *Renewable and Sustainable Energy Reviews* 82 (2018), pp. 2440–2454. ISSN: 13640321. DOI: 10.1016/j.rser.2017.09.003.

[190] Hiroshi Ito, Naoki Miyazaki, Masayoshi Ishida, and Akihiro Nakano. "Efficiency of unitized reversible fuel cell systems". In: *International Journal of Hydrogen Energy* 41.13 (2016), pp. 5803–5815. ISSN: 03603199. DOI: 10.1016/j.ijhydene.2016.01.150.

[191] S. Della Villa and C. Koeneke. "A Historical and Current Perspective of the Availability and Reliability Performance of Heavy Duty Gas Turbines: Benchmarks and Expectations". In: *Proceedings of the ASME Turbo Expo 2010.* New York, NY: ASME, 2010, pp. 835–845. ISBN: 978-0-7918-4396-3. DOI: 10.1115/GT2010-23182.

[192]   Raphael Lechner, Nicholas O'Connell, Thorsten Meierhofer, and Markus Brautsch. *Entwicklung, Umsetzung und Bewertung optimierter Monitoring- , Betriebs- und Regelstrategien für Blockheizkraftwerke: Abschlussbericht. Bearbeitungszeitraum: 11/2014 12/2016 Aktenzeichen: SWD-10.08.18.7- 14.* Vol. F 3058. Forschungsinitiative Zukunft Bau. Stuttgart: Fraunhofer IRB Verlag, 2018. ISBN: 9783738800838. URL: http://www.irbnet.de/daten/rswb/17129003225.pdf.

[193]   W. Colella, B. James, J. Moton, G. Saur, and T. Ramdsden. *Techno- economic Analysis of PEM Electrolysis for Hydrogen Production: Electrolytic Hydrogen Production Workshop.* Ed. by National Renewable Energy Laboratory. 2014. URL: https://www.energy.gov/sites/prod/files/2014/08/f18/fcto_2014_electrolytic_h2_wkshp_colella1.pdf.

[194]   Christopher Yang and Joan M. Ogden. "Determining the lowest-cost hydrogen delivery mode". In: *International Journal of Hydrogen Energy* 32.2 (2007), pp. 268–286. ISSN: 03603199. DOI: 10.1016/j.ijhydene.2006.05.009.

[195]   Susan Schoenung. *Economic Analysis of Large-Scale Hydrogen Storage for Renewable Utility Applications.* 2011. URL: https://prod-ng.sandia.gov/techlib-noauth/access-control.cgi/2011/114845.pdf.

[196]   Brian David James, Cassidy Houchins, Jennie Moton Huya-Kouadio, and Daniel A. DeSantis. *Final Report: Hydrogen Storage System Cost Analysis.* 2016. DOI: 10.2172/1343975.

[197]   Tom Smolinka, Martin Günther, and Jürgen Garche. *Stand und Entwicklungspotenzial der Wasserelektrolyse zur Herstellung von Wasserstoff aus regenerativen Energien : NOW-Studie : Kurzfassung des Abschlussberichts.* [Freiburg im Breisgau], 2011.

[198] Karim Ghaib and Fatima-Zahrae Ben-Fares. "Power-to-Methane: A state-of-the-art review". In: *Renewable and Sustainable Energy Reviews* 81 (2018), pp. 433–446. ISSN: 13640321. DOI: 10.1016/j.rser.2017.08.004.

[199] S. G. Pavlostathis and E. Giraldo–Gomez. "Kinetics of anaerobic treatment: A critical review". In: *Critical Reviews in Environmental Control* 21.5-6 (1991), pp. 411–490. ISSN: 1040-838X. DOI: 10.1080/10643389109388424.

[200] Rudolf K. Thauer, Anne-Kristin Kaster, Henning Seedorf, Wolfgang Buckel, and Reiner Hedderich. "Methanogenic archaea: ecologically relevant differences in energy conservation". In: *Nature reviews. Microbiology* 6.8 (2008), pp. 579–591. DOI: 10.1038/nrmicro1931.

[201] Anand B. Rao and Edward S. Rubin. "A technical, economic, and environmental assessment of amine-based CO2 capture technology for power plant greenhouse gas control". In: *Environmental science & technology* 36.20 (2002), pp. 4467–4475. DOI: 10.1021/es0158861.

[202] Jordi Guilera, Joan Ramon Morante, and Teresa Andreu. "Economic viability of SNG production from power and CO2". In: *Energy Conversion and Management* 162 (2018), pp. 218–224. ISSN: 01968904. DOI: 10.1016/j.enconman.2018.02.037.

[203] Jaques de Bucy. *The potential of power-to-gas: Technology review and economicpotential assessment.* Ed. by ENEA Consulting. Paris, France. URL: http://www.enea-consulting.com/wp-content/uploads/2016/01/ENEA-Consulting-The-potential-of-power-to-gas.pdf.

[204] Lukas Grond, Paula Schulze, and Johan Holstein. *System analyses Power to Gas: A technology review: Part of TKI project TKIG01038 – Systems analyses Power-to-Gas pathways Deliverable 1: Technology Review.* Groningen, 2013.

[205]   Stefan Schütz, Charlotte Große, and Cindy Kleinickel. *Treibhausgas-Minderungspotenziale in der europäischen Gasinfrastruktur: Im Auftrag des Umweltbundesamtes.* Leipzig, Freiberg. URL: https://www.bmu.de/fileadmin/Daten_BMU/Pools/Forschungsdatenbank/fkz_3712_41_344_treibhausgas_minderungspotenziale_bf.pdf.

[206]   BDEW. *Erdgasaufkommen und -verbrauch in Deutschland.* 2019. URL: https://www.bdew.de/service/daten-und-grafiken/erdgasaufkommen-und-verbrauch-deutschland/.

[207]   EWE NETZ GmbH, ed. *Netzentgelte Gas 2020: Örtliches Verteilnetz.* 2020. URL: https://www.ewe-netz.de/~/media/ewe-netz/downloads/2020_01_07_ewe_netz_nne_gas_oevn_2020.pdf.

[208]   Heizungsdiscount 24 GmbH, Stolzenmorgen 15, 35394 Gießen/Deutschland, Tel. 0641 / 948 252 00, www.heizungsdiscount24.de. *Buderus GB172 20 kW Logamax plus Gas-Brennwerttherme, E / H - Heizung und Solar zu Discountpreisen.* 27.05.2020. URL: https://www.heizungsdiscount24.de/gas-heizung/buderus-gb172-20-kw-logamax-plus-gas-brennwerttherme-e-h.html?gclid=EAIaIQobChMI16LCweTR6QIVlO7tCh1PzwJxEAQYBiABEgK1yvD_BwE.

[209]   Energy Consulting, Gesellschaft für Energiemanagement. "Kennziffernkatalog: Invesitionsvorbereitung in der Energiewirtschaft". In: (2004). URL: https://sisis.rz.htw-berlin.de/inh2011/12397792.pdf.

[210]   VDI 2067-1. *Wirtschaftlichkeit gebäudetechnischer Anlagen Grundlagen und Kostenberechnung.* 9.2012.

[211]   Vaillant GmbH. *Luft/Wasser-Wärmepumpe aroTHERM plus | Vaillant: COP über Temperaturdifferenz.* 26.05.2020. URL: https://www.

vaillant.de/heizung/produkte/luft-wasser-warmepumpe-arotherm-plus-194816.html.

[212] Heizungsdiscount 24 GmbH, Stolzenmorgen 15, 35394 Gießen/Deutschland, Tel. 0641 / 948 252 00, www.heizungsdiscount24.de. *Vaillant Durchlauferhitzer electronic VED E 27/7 Elektro-Durchlauferhitzer 27 kW - Heizung und Solar zu Discountpreisen*. 26.05.2020. URL: https://www.heizungsdiscount24.de/durchlauferhitzer/vaillant-durchlauferhitzer-electronic-ved-e-277-elektro-durchlauferhitzer-27-kw.html.

[213] Vaillant GmbH. *Vaillant Paket Luft/Wasser-Wärmepumpe aroTHERM Split VWL 35/5 AS - VWL 125/5 AS mit VIH RW 300/3 4.019-4.023, 4.126-4.130 | UNIDOMO: Produktbeschreibung und Kosten*. 26.05.2020. URL: https://www.unidomo.de/vaillant-paket-4-019-4-023-4-126-4-130-luft-wasser-waermepumpe-arotherm-split-vwl-35-5-as-vwl-125-5-as-mit-vih-rw-300-3-br_180485_161090/.

[214] Manuela Bücken et al. *Potenziale der Sektorkopplung und Nutzung von Strom aus Erneuerbaren Energien im Wärmebereich in Sachsen-Anhalt: Endbericht*. Ed. by EEB ENERKO GmbH and Mitteldeutsche Netzgesellschaft Strom. URL: https://enerko.de/wp-content/uploads/2017/12/Endbericht_PtH_web.pdf.

[215] Agora Energiewende, ed. *Power-to-Heat zur Integration von ansonsten abgeregeltem Strom aus Erneuerbaren Energien: Handlungsvorschläge basierend auf einer Analyse von Potenzialen und energiewirtschaftlichen Effekten*. URL: https://www.agora-energiewende.de/fileadmin2/Projekte/2013/power-to-heat/Agora_PtH_Langfassung_WEB.pdf.

[216] Ran Fu, Timothy Remo, and Robert M. Margolis. *2018 U.S. Utility-Scale Photovoltaics-Plus-Energy Storage System Costs Benchmark*. URL:

https://www.nrel.gov/docs/fy19osti/71714.
pdf.

[217] Wesley Cole and A. Will Frazier. *Cost Projections for Utility-Scale Battery Storage*. URL: https://www.nrel.gov/docs/fy19osti/73222.pdf.

[218] Eduardo Redondo-Iglesias, Pascal Venet, and Serge Pelissier. "Global Model for Self-Discharge and Capacity Fade in Lithium-Ion Batteries Based on the Generalized Eyring Relationship". In: *IEEE Transactions on Vehicular Technology* 67.1 (2018), pp. 104–113. ISSN: 0018-9545. DOI: 10.1109/TVT.2017.2751218. URL: https://hal.archives-ouvertes.fr/hal-01593047/document.

[219] IRENA. *Electricity storage and renewables: Costs and markets to 2030*. Abu Dhabi, 2017.

[220] Jens Noack, Lars Wietschel, Nataliya Roznyatovskaya, Karsten Pinkwart, and Jens Tübke. "Techno-Economic Modeling and Analysis of Redox Flow Battery Systems". In: *Energies* 9.8 (2016), p. 627. ISSN: 1996-1073. DOI: 10.3390/en9080627.

[221] CellCube Energy Storage Systems Inc., ed. *CellCube Investor Presentation*. URL: https://static1.squarespace.com/static/5b1198ada2772c6585959926/t/5c9cd83ff9619a021c2de4c4/1553782876741/CellCube+Investor+Presentation+-+2019+March+26+%281%29.pdf.

[222] Jens F. Peters, Jacob Fulton, Manuel J. Baumann, and Marcel Weil. "Life Cycle Costs Model for Vanadium Redox Flow Batteries: Poster at 11th Society and Materials International Conference (SAM 11)". In: (2017). URL: https://www.researchgate.net/publication/317249223_Life_Cycle_Costs_Model_for_Vanadium_Redox_Flow_Batteries.

[223]   Gerrit Erichsen, Ball Chrstopher, Kather Alfons, and Wilhelm Kuckshinrichs. *Data Documentation: VEREKON Cost Paremters for 2050 in its Energy System Model.* 2019. URL: `https://tore.tuhh.de/bitstream/11420/4409/2/documentation_verekon_cost_parameters_for_2050_in_its_energy_system_model.pdf`.

[224]   Volterion GmbH, ed. *Volterion - power RFB: Technical Specifications.* URL: `https://www.volterion.com/wp-content/uploads/2019/03/VOLT_powerRFB_Productsheet.pdf`.

[225]   Landesamt für Bergbau, Energie und Geologie. *Erdöl und Erdgas in der Bundesrepublik Deutschland 2019.* Hannover, 2020. URL: `https://www.lbeg.niedersachsen.de/erdoel-erdgas-jahresbericht/jahresbericht-erdoel-und-erdgas-in-der-bundesrepublik-deutschland-936.html`.

[226]   Sabine Donadei and Gregor-Sönke Schneider. "Wasserstoffspeicherung in Salzkavernen". In: *Wasserstoff und Brennstoffzelle.* Ed. by Johannes Töpler and Jochen Lehmann. Berlin and Heidelberg: Springer Vieweg, 2017, pp. 315–325. ISBN: 9783662533598. DOI: `10.1007/978-3-662-53360-4{\textunderscore}15`. URL: `https://link.springer.com/content/pdf/10.1007%2F978-3-662-53360-4_15.pdf`.

[227]   Xing Ju, Chao Xu, Xue Han, Hui Zhang, Gaosheng Wei, and Lin Chen. "Recent advances in the PV-CSP hybrid solar power technology". In: AIP Conference Proceedings. Author(s), 2017, p. 110006. DOI: `10.1063/1.4984480`. URL: `https://aip.scitation.org/doi/pdf/10.1063/1.4984480`.

[228]   A. Green, C. Diep, R. Dunn, and J. Dent. "High Capacity Factor CSP-PV Hybrid Systems". In: *Energy Procedia* 69 (2015), pp. 2049–2059. ISSN: 18766102. DOI: `10.1016/j.egypro.2015.03.218`.

[229] E. Casati, F. Casella, and P. Colonna. "Design of CSP plants with opti-
        mally operated thermal storage". In: *Solar Energy* 116 (2015), pp. 371–
        387. ISSN: 0038092X. DOI: 10.1016/j.solener.2015.03.
        048.

[230] G. NREL Glatzmaier. *Developing a Cost Model and Methodology to
        Estimate Capital Costs for Thermal Energy Storage.* Ed. by National
        Renewable Energy Laboratory. 2011. URL: https://www.nrel.
        gov/docs/fy12osti/53066.pdf.

[231] Bundesverband Energiespeicher, ed. *Fact Sheet Speichertechnologien:
        Hochtemperatur Flüssigspeicher.* 2017. URL: https://www.solarthermalwor.
        org/sites/default/files/news/file/2018-03-
        11/bves_moltan_salt_storage.pdf.

[232] Gregory Wilk, Alfred DeAngelis, and Asegun Henry. "Estimating the
        cost of high temperature liquid metal based concentrated solar power".
        In: *Journal of Renewable and Sustainable Energy* 10.2 (2018), p. 023705.
        DOI: 10.1063/1.5014054.

[233] Heizungsdiscount 24 GmbH. *1000 Liter TWL Puffer, Pufferspeicher Typ P
        1000, Iso-C - Heizung und Solar zu Discountpreisen.* 2020. URL: https:
        //www.heizungsdiscount24.de/speichertechnik/
        1000-liter-twl-puffer-pufferspeicher-typ-
        p-1000-iso-c.html?gclid=EAIaIQobChMImZSZ-
        __i6QIVAWHmCh32WwHSEAQYASABEgJxPPD_BwE.

[234] M. Pehnt. *Wärmenetzsysteme 4.0: Endbericht – Kurzstudie zur Umset-
        zung der Maßnahme „Modellvorhaben erneuerbare Energien in hochef-
        fizienten Niedertemperaturwärmenetzen".* Ed. by ifeu, adelphi, Eco-
        fys, PwC, dena, AEE. Heidelberg, Berlin, Düsseldorf, Köln, 2017. URL:
        https://www.dena.de/fileadmin/dena/Dokumente/
        Themen_und_Projekte/Gebaeude/Rahmenvertrag_
        BMWi/Studie_Umsetzung_Modellvorhaben_erneuerbare_

Energien_hocheffiziente_saisonalspeichergestuetz
Niedertemperaturwaermenetze.pdf.

[235] TWL-Technologie GmbH. *ÖkoLine-Profi Isolierungen: ERP-Ökoline C - 1000 Liter*. 2020. URL: https : / / twl - technologie . de / files/6185/upload/kunden/kundenbereich/login_ datein / datenblaetter / speicherzubehoer / neu / info/02_O%CC%88koLine-Profi.pdf.

[236] YCharts Inc. *European Union Natural Gas Import Price*. 28.10.2022. URL: https : / / ycharts . com / indicators / europe_ natural_gas_price.

[237] Hans-Dieter Karl. "Abschätzung der Förderkosten für Energierohstoffe". In: *ifo Schnelldienst* 63.2 (2010), pp. 21–29. ISSN: 0018-974 X.

[238] Bundesumweltamt. *Methodenkonvention 3.0 zur Ermittlung von Umweltkosten - Kostensätze*. 2018. URL: https : / / www . umweltbundesamt . de / sites / default / files / medien / 384 / bilder / dateien / 3 _ tab _ uba - empfehlung - klimakosten_ 2019-01-17.pdf.

[239] NREL. *Data | Electricity | ATB | NREL*. 2020. URL: https : / / atb . nrel.gov/electricity/2020/data.php.

[240] Dongkuan Xu and Yingjie Tian. "A Comprehensive Survey of Clustering Algorithms". In: *Annals of Data Science* 2.2 (2015), pp. 165–193. ISSN: 2198-5812. DOI: 10 . 1007 / s40745 - 015 - 0040 - 1.

[241] Ricardo J. G. B. Campello, Davoud Moulavi, and Joerg Sander. "Density-Based Clustering Based on Hierarchical Density Estimates". In: *Advances in Knowledge Discovery and Data Mining*. Ed. by David Hutchison et al. Vol. 7819. Lecture Notes in Computer Science. Berlin, Heidelberg: Springer Berlin Heidelberg, 2013, pp. 160–172. ISBN: 978-3-642-37455-5. DOI: 10 . 1007 / 978 - 3 - 642 - 37456 - 2 { \textundersco 14.

[242] Leland McInnes, John Healy, and Steve Astels. "hdbscan: Hierarchical density based clustering". In: *The Journal of Open Source Software* 2.11 (2017), p. 205. DOI: 10.21105/joss.00205.

[243] Nikolay Oskolkov. *How to cluster in High Dimensions | by Nikolay Oskolkov | Towards Data Science.* 2019. URL: https://towardsdatascience.com/how-to-cluster-in-high-dimensions-4ef693bacc6.

[244] Christoph Molnar. *Interpretable machine learning: A guide for making black box models explainable.* 1st edition. 2019. ISBN: 9780244768522.

[245] R. R. Selmic and F. L. Lewis. "Neural-network approximation of piecewise continuous functions: application to friction compensation". In: *IEEE transactions on neural networks* 13.3 (2002), pp. 745–751. ISSN: 1045-9227. DOI: 10.1109/TNN.2002.1000141.

[246] tisimst. *pyDOE.* 2017. URL: https://pythonhosted.org/pyDOE/#.

[247] The SciPy community. *scipy.stats.truncnorm.* 17.12.2018. URL: https://docs.scipy.org/doc/scipy/reference/generated/scipy.stats.truncnorm.html.

[248] Tyler Stehly and Philipp Beiter. *2018 Cost of Wind Energy Review.* 2019. URL: https://www.nrel.gov/docs/fy20osti/74598.pdf.

[249] Ran Fu, Timothy W. Remo, and Robert M. Margolis. *2018 U.S. Utility-Scale Photovoltaics-Plus-Energy Storage System Costs Benchmark.* 2018. DOI: 10.2172/1483474.

Appendix A

Investigation of Systematic Errors

The energy system models parameters are specified in table A.1 and table A.2 and are based on [95]. 2016 costs and efficiency extremes are based on literature to assure plausibility. As the exact values are not relevant for the findings of this study, costs and efficiencies are not referenced.

For each technology except for battery power and hydrogen gas turbines an input parameter distribution is defined. The distribution is based on a latin hyper cube over the whole considered parameter range. This range is from ten to 110 % of 2016 cost for all costs and from minimum to maximum efficiency for the efficiencies except battery round trip efficiency, which is assumed to be constant. The even distribution over each parameter range is then transformed in such a way, that the frequency represents the assumed probability distribution. The probability distributions and the frequencies of four different standard deviations are depicted in fig. A.1. For all distributions applied in this work the mean lies at 50 % of the considered range and the distribution is adjusted in such a way, that the distribution becomes zero at the edges of the parameter range. This leads to slightly lower standard deviations than those applied initially. The distribution is applied to the sum of annualized and fixed cost as well as to variable cost. In the main text, standard

Table A.1: Technologies and economic assumptions regarding the model energy system.

technology	cost$_{2016}$	depr.[a]	FOM in $€\,kW^{-1}\,a^{-1}$	annuity[b] in $€\,MWh^{-1}$	VOM in $€\,MWh^{-1}$
photovoltaics	900 $€\,kW_{peak}^{-1}$	25	-	66.61	-
onshore wind	2060 $€\,kW_{peak}^{-1}$	25	56	187.87	10
offshore wind	4500 $€\,kW_{peak}^{-1}$	25	136	424.05	15
AE	1500 $€\,kW^{-1}$	25	9	96.02	-
SOFC stack	1100 $€\,kW^{-1}$	10	66	204.22	6.78
SOFC non stack	1100 $€\,kW^{-1}$	25	-	-	
battery capacity	450 $€\,kWh^{-1}$	10	-	55.48	-
H_2 storage	0.0943 $€\,kWh^{-1}$	30	-	0.00569	-
water for H_2	0.363 $€\,kWh^{-1}$	25	-	-	
battery power	100 $€\,kW^{-1}$	25	5	11.4	-
H_2 gas turbine	700 $€\,kW^{-1}$	30	21	38.12	

[a]depreciation period in years
[b]annualized cost at 2 % interest and 2 % inflation

Table A.2: Efficiencies subject to uncertainty in the model energy system.

	Efficiency input distribution		
technology	max	min	mean
AE	0.75	0.532	0.641
SOFC	0.852	0.456	0.654

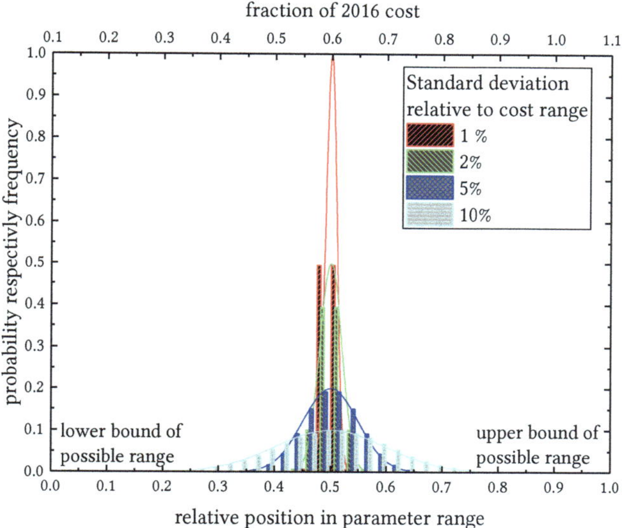

Figure A.1: Input distributions histograms and curves relative to the parameter ranges and 2016 cost.

deviations are given in percent of the considered parameter range. Standard deviations relative to the mean can be calculated according to eq. (A.1).

$$\sigma_{mean} = \sigma_{range} \frac{range}{mean * range} \tag{A.1}$$

For batteries 365 cycles per year and twenty percent degradation over 6000 cycles are assumed. Based on these assumptions the medium available capacity during 10 years of service is calculated to 93.65 %. The round trip efficiency of batteries is assumed to be 90 %.

A.1 Latin Hyper Cube and Transformation

The methodology of the use of various strategies to obtain multi-dimensional distributions which are favorable for the derivation of surrogate models from simulations is detailed in [103]. The basic idea is that sampling points should be selected which cover the whole parameter space while minimizing the variance of the parameters global mean prediction [103]. One of the methods which performs better than a Monte-Carlo distribution at producing low variance fields are latin hyper cubes [103]. This is especially relevant as Monte-Carlo works well if a very high number of samples has to be distributed which is not always feasible, considering the effort of an optimization. The latin hyper cube works by dividing the space into slices and then distributing the available samples in these slices. Herein the latin hyper cube generation method from the python package pyDOE [246] is used with a centered design. In order to achieve a distribution with defined standard deviation and mean to introduce expected probabilies the function *truncnorm* is used from the scipy.stats package [247].

A grafic represenation of a generated latin hyper cube with 50 samples and the result of the transformation is depicted in fig. A.2. The lhc is evenly distributed in both dimensions while at the same time covering a great part of the parameter space. With the transformation the lhc is compressed into the relevant space and the resulting distribution adheres to the predefined standard deviation and mean in both dimensions. The goal of this approach in this work is to achieve a set of samples which is well distributed in space while at the same time adhering to the constraints regarding standard deviation and position of the mean in all dimensions.

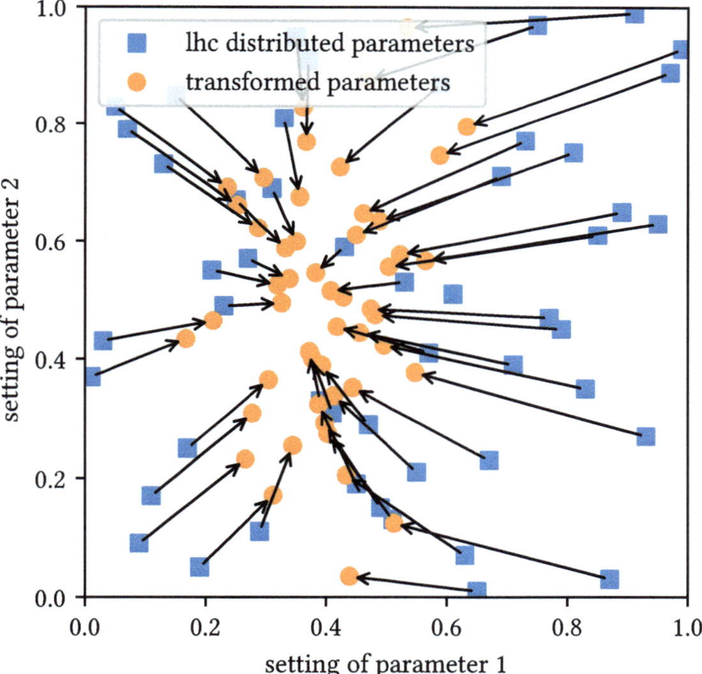

Figure A.2: The arrows show how each sample point is moved during the transformation. The mean of the tranformed distribution is (0.4, 0.5), the standard deviation is 0.1 for parameter 1 and 0.2 for parameter 2.

Appendix B

Energy System Opimization

B.1 Probable Cost Reductions

Table B.1: Photovoltaic component cost reductions and distributions until 2050.

Component	% of cost	Ref.	$\frac{2050_{cost}}{2017_{cost}}$	Ref.	Rel.SD	Ref.
Power Electronics	11%	[63]	39%	[63]*	44%	[63]*
Module	55%	[63]	54%	[63]*	55%	[63]*
BOS	34%	[63]	57%	[63]*	33%	[63]*

*scaled to fit the total cost reductions and SD of [239]

Table B.2: Onshore wind component cost reductions and distributions until 2050.

Component	% of cost	Ref.	$\frac{2050_{cost}}{2017_{cost}}$	Ref.	Rel.SD	Ref.
Soft Cost	6.2%	[248]	55.5%	*	46.4%	*
BOS	16.1%	[248]				a*
Turbine	39.0%	[248]	45.4%	a	46.4%	a*
Power Electr.	10.2%	[248]	39.3%	table B.1	44.1%	table B.1
O&M	28.5%	[239]	75.0%	[239]	13.9%	[239]

13 % of [248] turbine cost is accounted for as power electronics
*scaled to fit the total cost reductions and SD of [239]
^{a}cost red. or SD of complete system according to table 5.19 for component

Table B.3: Offshore wind component cost reductions and distributions until 2050.

Component	% of cost	Ref.	$\frac{2050_{cost}}{2017_{cost}}$	Ref.	Rel.SD	Ref.
Soft Cost	9.6%	[248]	18.3%	*	21.7%	a*
BOS	37.1%	[248]				
Turbine	15.3%	[248]	45.4%	table B.2	46.4%	table B.2
Power Electr.	4.0%	[248]	39.3%	table B.1	44.1%	table B.1
O&M	34.0%	[248]	29.5%	[239]	10.2%	[239]

13 % of [248] turbine cost is accounted for as power electronics
*scaled to fit the total cost reductions and SD of [239]
^{a} cost red. or SD of complete system according to table 5.19 for component

Table B.4: Battery capacity component cost reductions and distributions until 2050.

Component	% of cost	Ref.	$\frac{2050_{cost}}{2017_{cost}}$	Ref.	Rel.SD	Ref.
Cell	73.8%	[249]	41.2%	[a]	29.2%	[a*]
BOS	15.0%	[249]				
Developer Cost	11.3%	[249]	41.2%	[a]	29.2%	[a*]

*scaled to fit the total cost reductions and SD of [239]
[a]cost red. or SD of complete system according to table 5.19 for component

Table B.5: Battery power component cost reductions and distributions until 2050.

Component	% of cost	Ref.	$\frac{2050_{cost}}{2017_{cost}}$	Ref.	Rel.SD	Ref.
Power Electr.	25.4%	[249]	39.3%	table B.1	44.1%	table B.1
BOS	49.6%	[249]	41.5% *		45.3%	*
Developer Cost	25.0%	[249]	41.2%	table B.4	29.2%	table B.4

*scaled to fit the total cost reductions and SD of [239]

Table B.6: PEM component cost reductions and distributions until 2050.

Component	% of cost	Ref.	$\frac{2050_{cost}}{2017_{cost}}$	Ref.	Rel.SD	Ref.
Stack	25.5%	[188]	21.3%	*	23.7%	[a*]
Gas Conditioning	22.5%	[188]				
Balance of Plant	11.3%	[188]				
Installation	25.0%	[193]				
Power Electr.	15.8%	[188]	39.3%	table B.1	44.1%	table B.1

*scaled to fit the total cost reductions and SD of [239]
[a]cost red. or SD of complete system according to table 5.19 for component

Table B.7: AEL component cost reductions and distributions until 2050.

Component	% of cost	Ref.	$\frac{2050_{cost}}{2017_{cost}}$	Ref.	Rel.SD	Ref.
Stack	32.3%	[188]				
Gas Conditioning	20.3%	[188]	40.1%	*	22.2%	a*
Balance of Plant	12.8%	[188]				
Installation	25.0%	a.a.				
Power Electr.	9.8%	[188]	39.3%	table B.1	44.1%	table B.1

*scaled to fit the total cost reductions and SD of [239]
[a] cost red. or SD of complete system according to table 5.19 for component

Table B.8: Redox flow battery power component cost reductions and distributions until 2050.

Component	% of cost	Ref.	$\frac{2050_{cost}}{2017_{cost}}$	Ref.	Rel.SD	Ref.
Stack	45.5%	[220]				
Other	20.5%	[220]	53.1%	*	21.6%	a*
Installation	25.0%	a.a				
Power Electr.	9.1%	[220]	39.3%	table B.1	44.1%	table B.1

a.a.: author's assumption
*scaled to fit the total cost reductions and SD of [239]
[a] cost red. or SD of complete system according to table 5.19 for component

Appendix C

Results

C.1 Stochastic Optimization Results at 95 % Emission Reduction

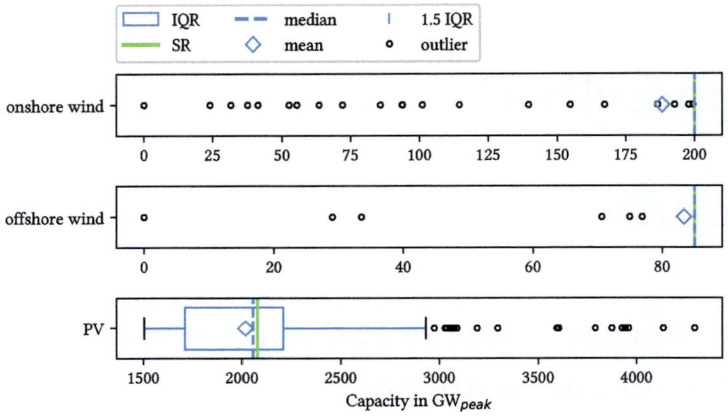

Figure C.1: Stochastic optimization result for renewable electricity production technologies in the 95 % emission reduction case. Technologies not displayed are not installed in any of the cases.

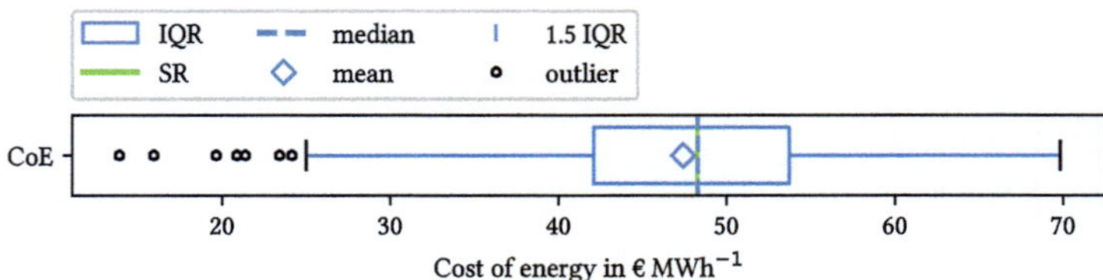

Figure C.2: Stochastic optimization result for the cost of energy in the 95 % emission reduction case.

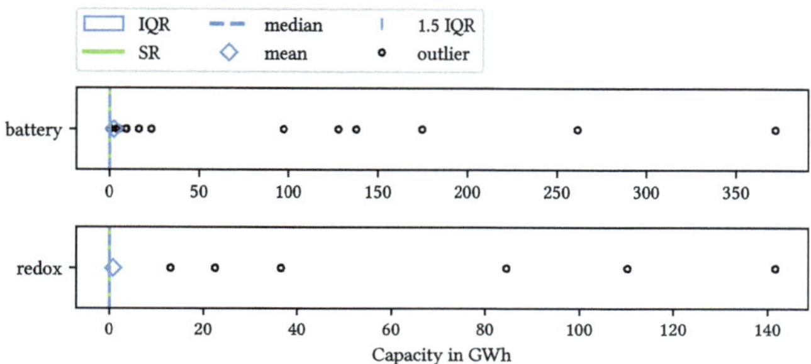

Figure C.3: Stochastic optimization result for electricity storage in the 95 % emission reduction case. Technologies not displayed are not installed in any of the cases.

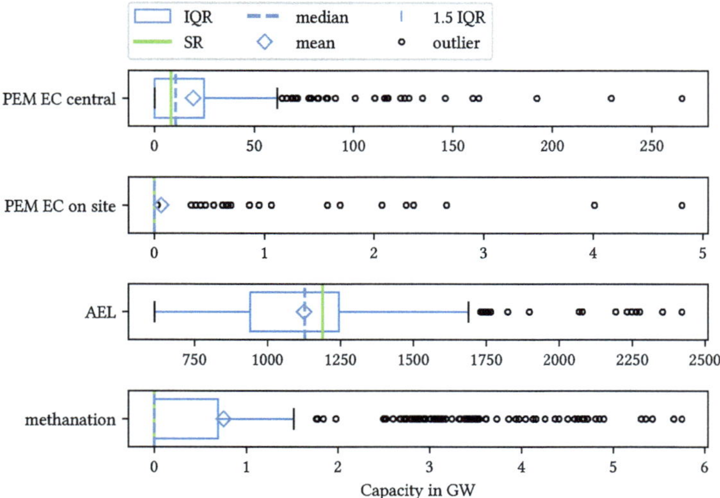

Figure C.4: Stochastic optimization result for power to gas technologies in the 95 % emission reduction case. Technologies not displayed are not installed in any of the cases.

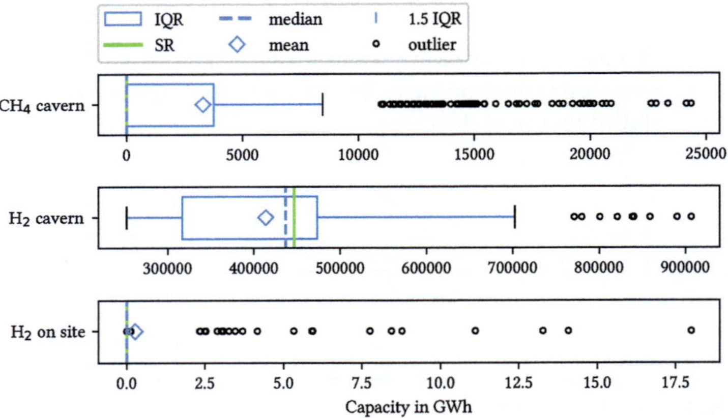

Figure C.5: Stochastic optimization result for gas storage in the 95 % emission reduction case. Technologies not displayed are not installed in any of the cases.

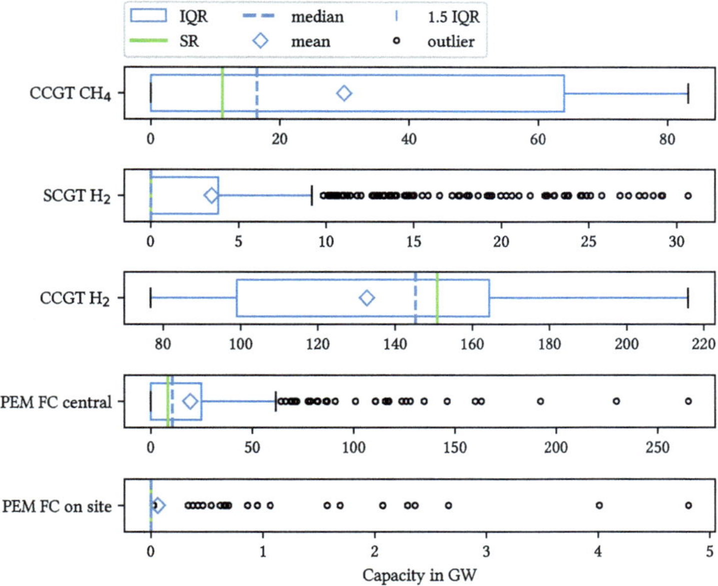

Figure C.6: Stochastic optimization result for electricity generation from gas in the 95 % emission reduction case. Technologies not displayed are not installed in any of the cases.

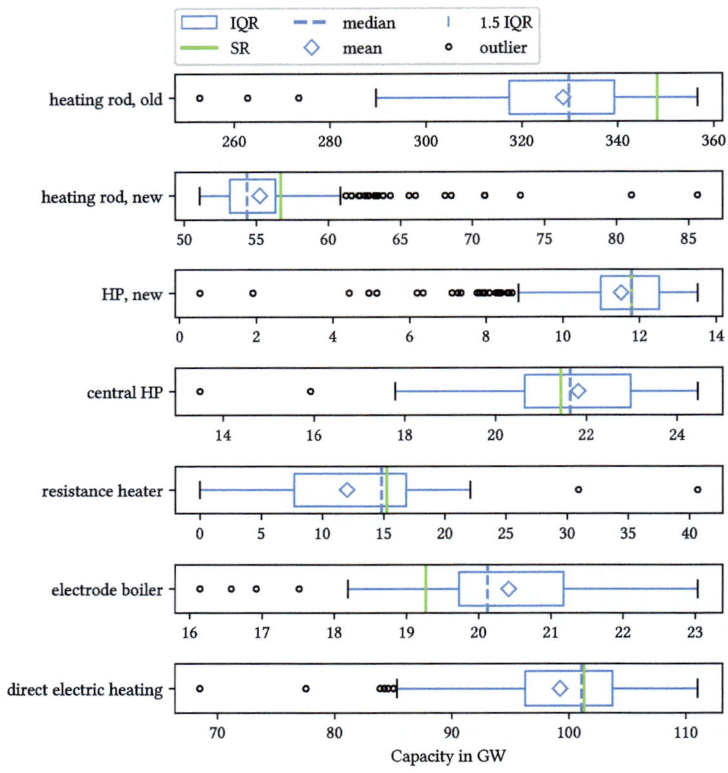

Figure C.7: Stochastic optimization result for electrical heating technologies in the 95 % emission reduction case. Technologies not displayed are not installed in any of the cases.

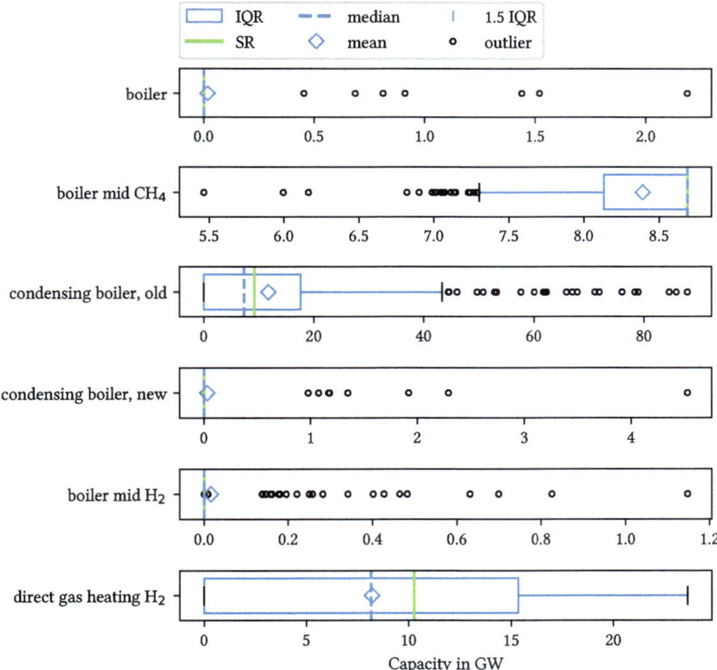

Figure C.8: Stochastic optimization result for heating with gaseous energy carriers in the 95 % emission reduction case. Technologies not displayed are not installed in any of the cases.

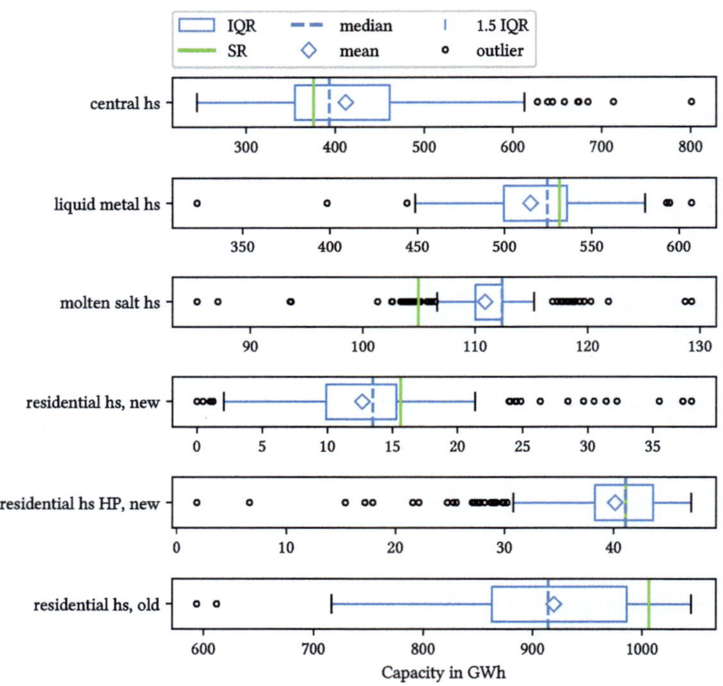

Figure C.9: Stochastic optimization result for heat storage in the 95 % emission reduction case. Technologies not displayed are not installed in any of the cases.

C.2 Stochastic Optimization Results at 80 % Emission Reduction

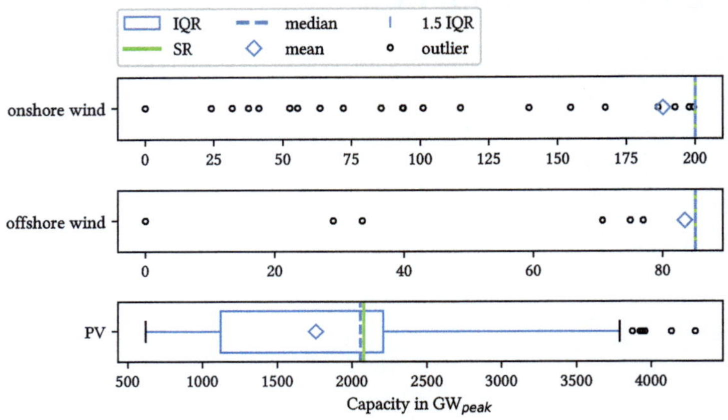

Figure C.10: Stochastic optimization result for renewable electricity production technologies in the 80 % emission reduction case. Technologies not displayed are not installed in any of the cases.

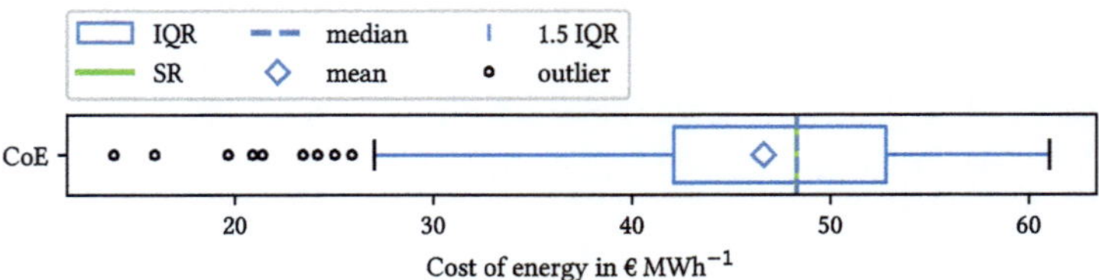

Figure C.11: Stochastic optimization result for the cost of energy in the 80 % emission reduction case.

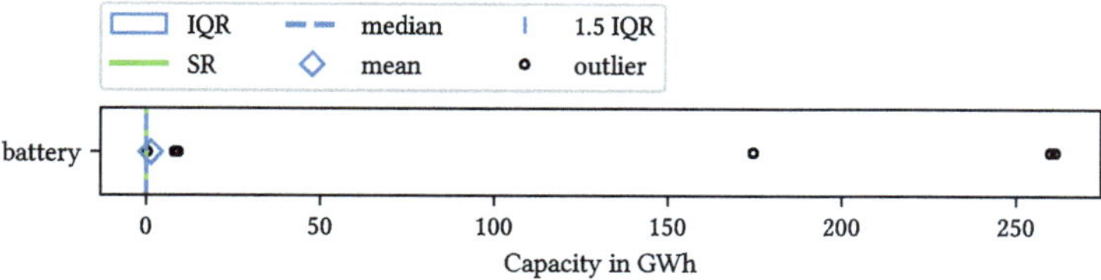

Figure C.12: Stochastic optimization result for electricity storage in the 80 % emission reduction case. Technologies not displayed are not installed in any of the cases.

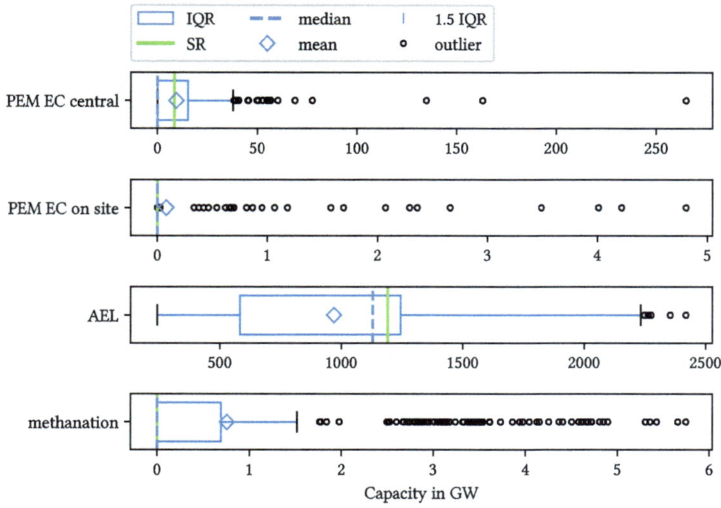

Figure C.13: Stochastic optimization result for power to gas technologies in the 80 % emission reduction case. Technologies not displayed are not installed in any of the cases.

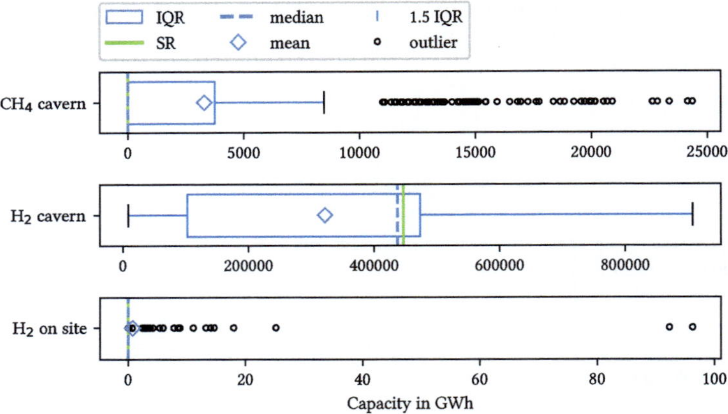

Figure C.14: Stochastic optimization result for gas storage in the 80 % emission reduction case. Technologies not displayed are not installed in any of the cases.

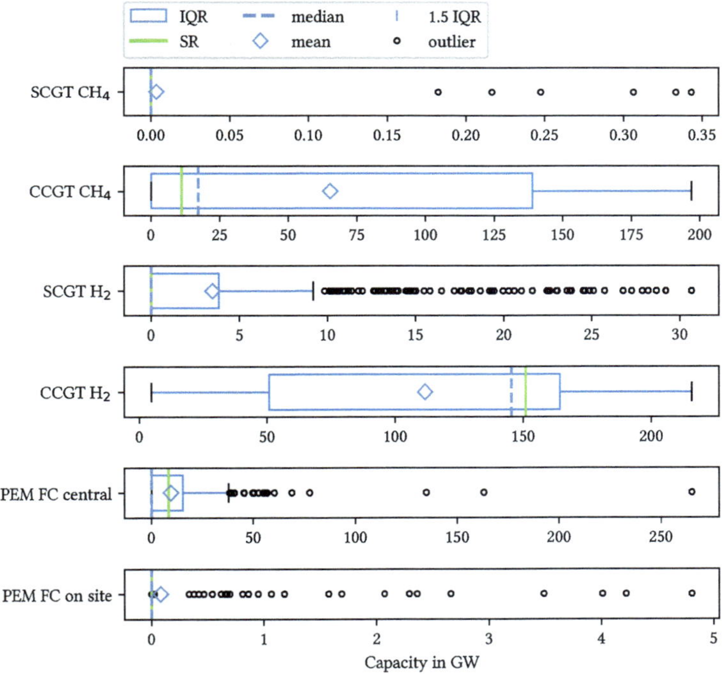

Figure C.15: Stochastic optimization result for electricity generation from gas in the 80 % emission reduction case. Technologies not displayed are not installed in any of the cases.

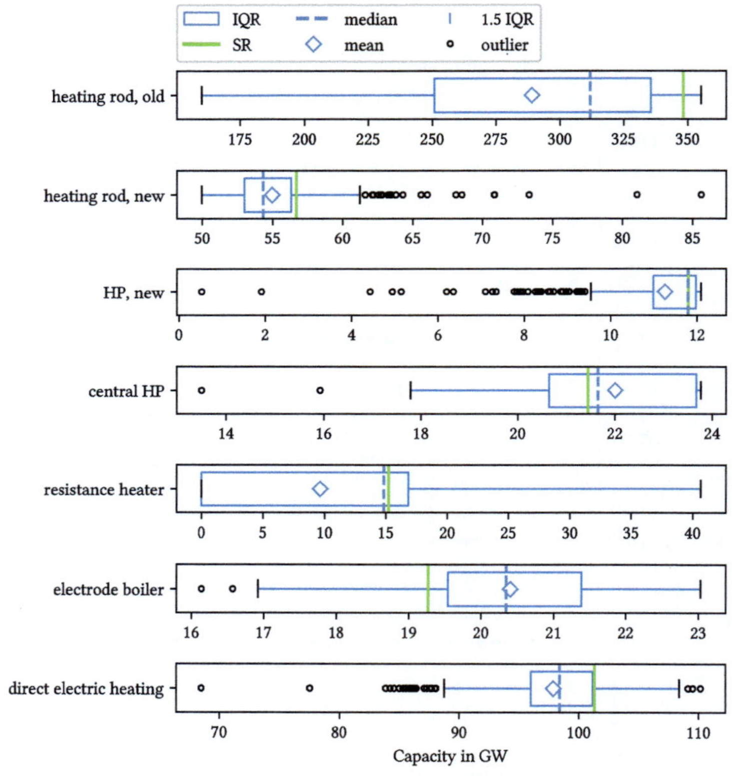

Figure C.16: Stochastic optimization result for electrical heating technologies in the 80 % emission reduction case. Technologies not displayed are not installed in any of the cases.

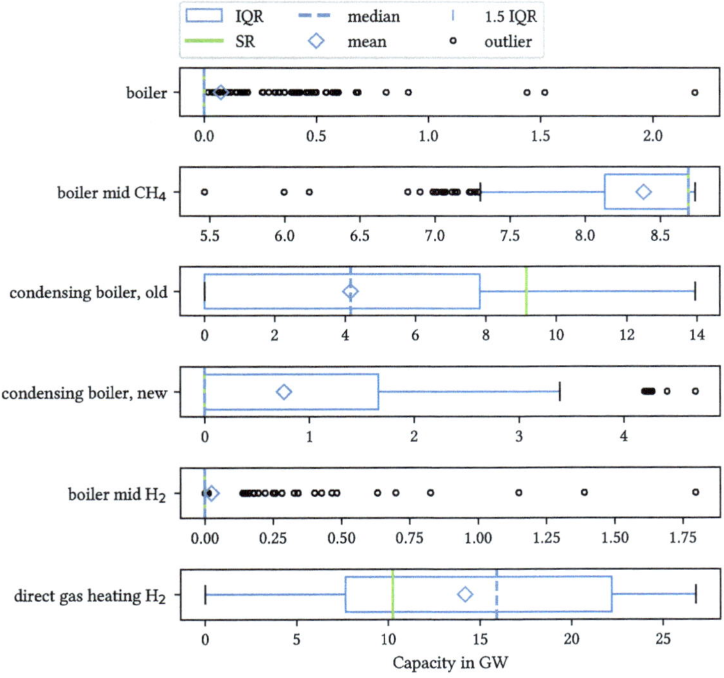

Figure C.17: Stochastic optimization result for heating with gaseous energy carriers in the 80 % emission reduction case. Technologies not displayed are not installed in any of the cases.

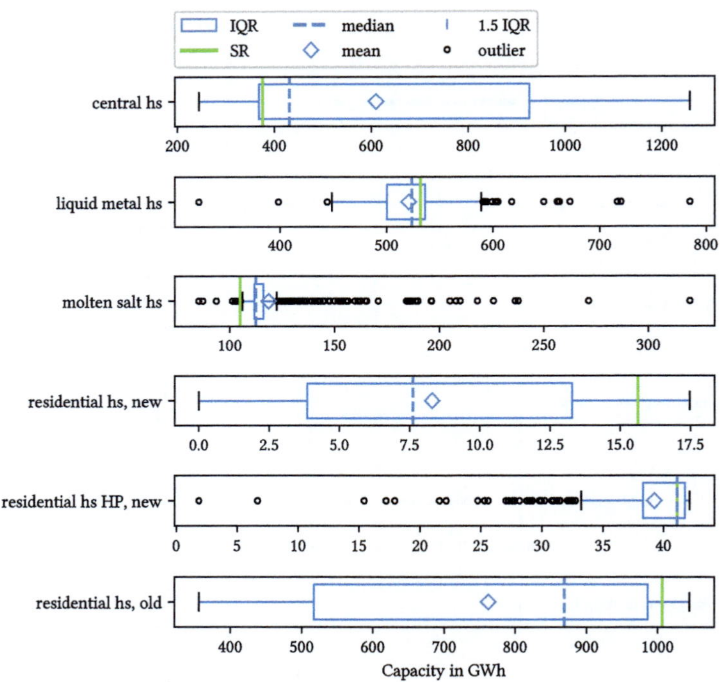

Figure C.18: Stochastic optimization result for heat storage in the 80 % emission reduction case. Technologies not displayed are not installed in any of the cases.

C.3 Patterns within the optimization results - 95% emission reduction

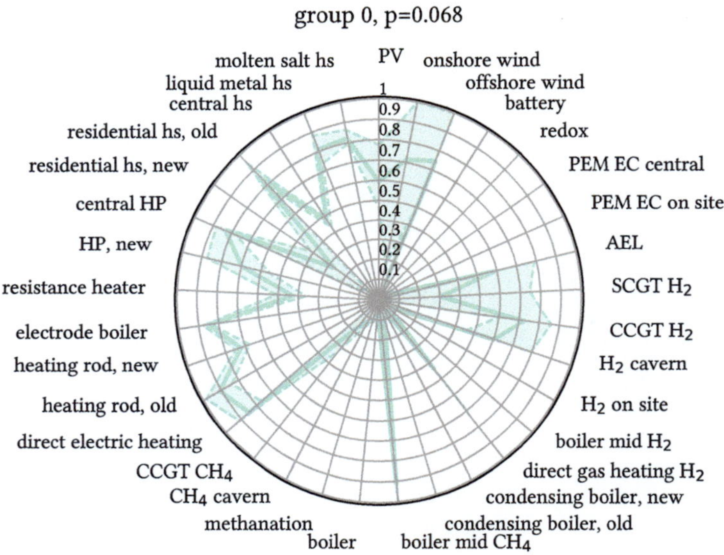

Figure C.19: Clusters resulting from the clustering of the optimization results under the 95 % carbon emission reduction boundary condition. The radial axis displays the relative amount of optimized technology in the range between zero and maximal installed capacity of each technology in the given emission scenario, the solid line shows the mean values, dashed lines depicts min and max of the IQR, which is colored.

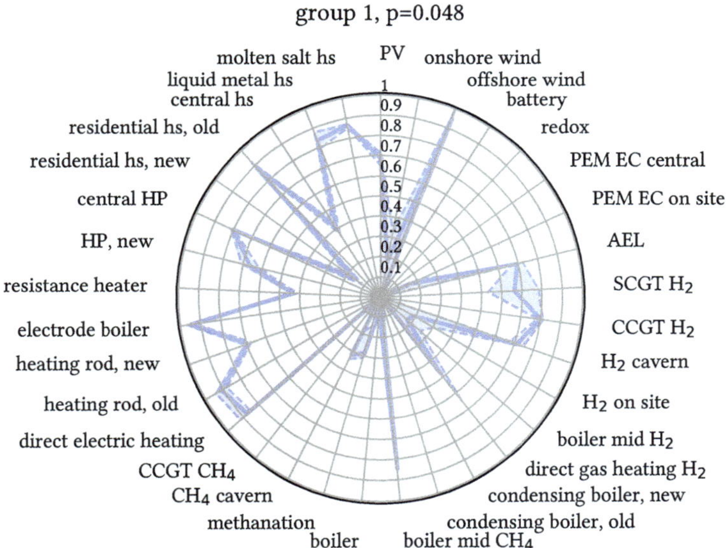

Figure C.20: Clusters resulting from the clustering of the optimization results under the 95 % carbon emission reduction boundary condition. The radial axis displays the relative amount of optimized technology in the range between zero and maximal installed capacity of each technology in the given emission scenario, the solid line shows the mean values, dashed lines depicts min and max of the IQR, which is colored.

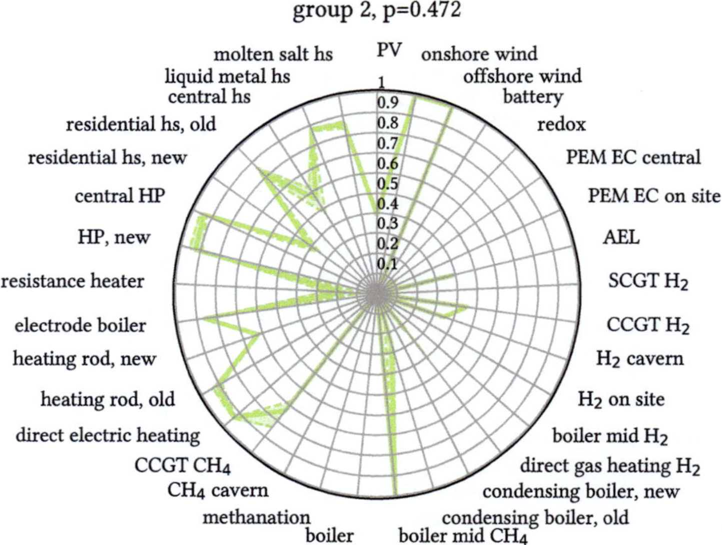

Figure C.21: Clusters resulting from the clustering of the optimization results under the 95 % carbon emission reduction boundary condition. The radial axis displays the relative amount of optimized technology in the range between zero and maximal installed capacity of each technology in the given emission scenario, the solid line shows the mean values, dashed lines depicts min and max of the IQR, which is colored.

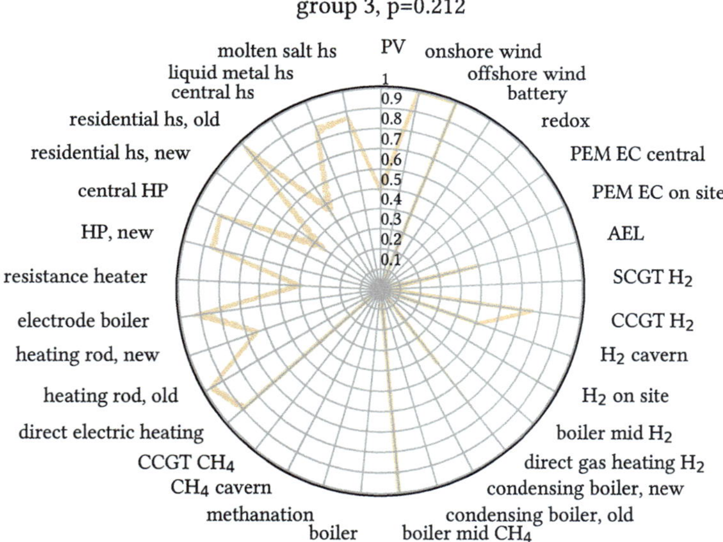

Figure C.22: Clusters resulting from the clustering of the optimization results under the 95 % carbon emission reduction boundary condition. The radial axis displays the relative amount of optimized technology in the range between zero and maximal installed capacity of each technology in the given emission scenario, the solid line shows the mean values, dashed lines depicts min and max of the IQR, which is colored.

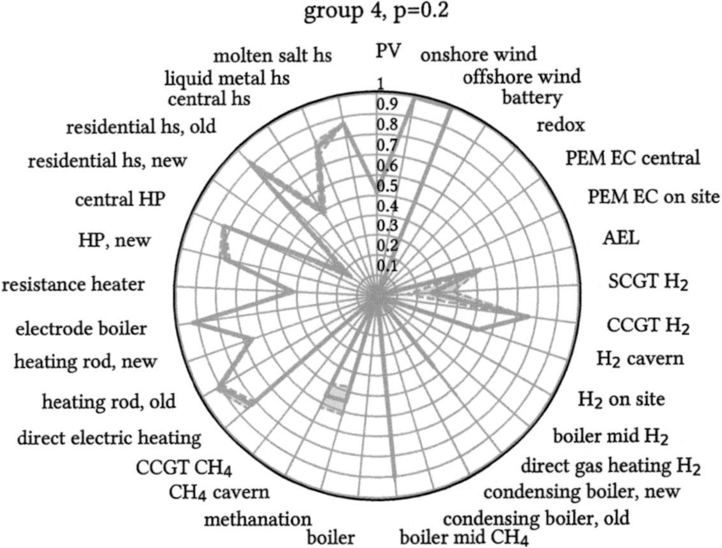

Figure C.23: Clusters resulting from the clustering of the optimization results under the 95 % carbon emission reduction boundary condition. The radial axis displays the relative amount of optimized technology in the range between zero and maximal installed capacity of each technology in the given emission scenario, the solid line shows the mean values, dashed lines depicts min and max of the IQR, which is colored.

Table C.1: Relative cost mean of input parameters, carbon emissions in million tons and cost of energy in € MWh^{-1} by cluster in case of 95% emission reduction.

	all cases	gr. 0	gr. 1	gr. 2	gr. 3	gr. 4
probability:		0.08	0.048	0.46	0.21	0.202
photovoltaic	0.53	0.39	0.25	0.67	0.48	0.36
onshore wind	0.55	0.52	0.63	0.55	0.55	0.55
offshore wind	0.27	0.26	0.29	0.27	0.27	0.27
battery, capacity	0.41	0.39	0.41	0.42	0.42	0.40
battery, power	0.41	0.41	0.39	0.41	0.42	0.39
PEM	0.24	0.23	0.24	0.24	0.24	0.24
alk. electrolysis	0.40	0.40	0.39	0.40	0.39	0.40
methanation	0.47	0.49	0.47	0.46	0.48	0.46
redox power	0.52	0.50	0.48	0.52	0.53	0.51
redox capacity	0.24	0.25	0.22	0.24	0.24	0.24
molten strorage	0.29	0.27	0.28	0.30	0.30	0.29
cost of energy	47.4	38.8	33.3	54.4	45.8	39.0
CO$_2$ emissions	24.9	11.2	0	48.6	5.71	0

C.4 Patterns within the optimization results - 80 % emission reduction

C.5 Clustering comparison of benchmark and reduced demand stochastic analysis

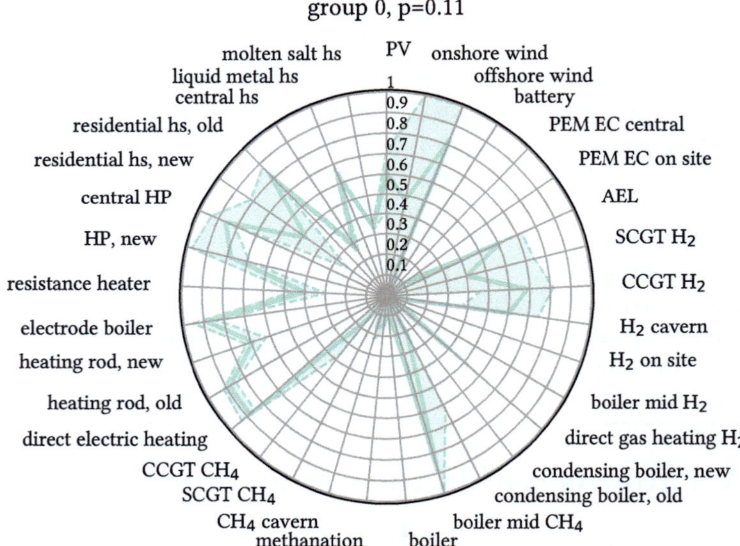

Figure C.24: Group 0 resulting from the clustering of the stochastic optimization in the 80 % emission reduction case. The radial axis displays the relative amount of optimized technology in the range between zero and maximal installed capacity of each technology in the given emission scenario, the solid line shows the mean values, dashed lines depicts min and max of the IQR, which is colored.

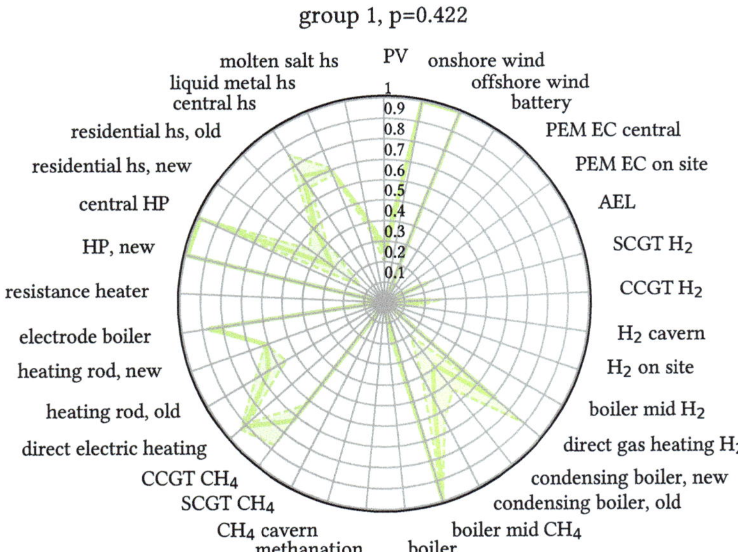

Figure C.25: Cluster resulting from the clustering of the stochastic optimization in the 80 % emission reduction case. The radial axis displays the relative amount of optimized technology in the range between zero and maximal installed capacity of each technology in the given emission scenario, the solid line shows the mean values, dashed lines depicts min and max of the IQR, which is colored.

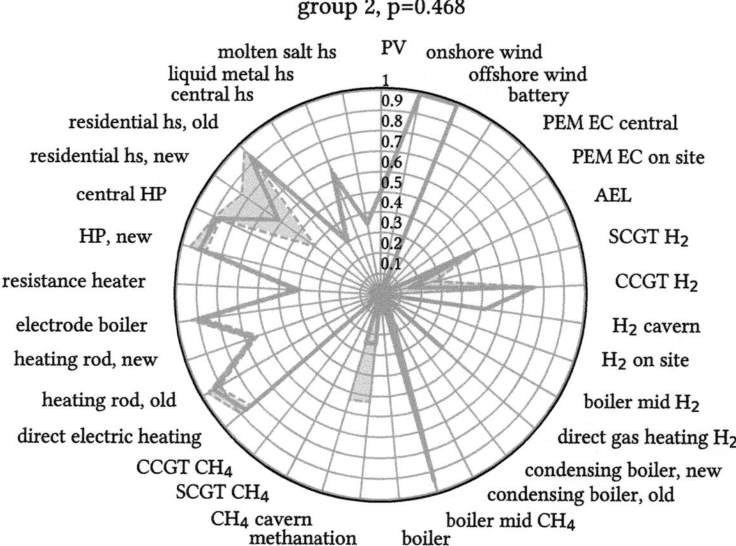

Figure C.26: Cluster resulting from the clustering of the stochastic optimization in the 80 % emission reduction case. The radial axis displays the relative amount of optimized technology in the range between zero and maximal installed capacity of each technology in the given emission scenario, the solid line shows the mean values, dashed lines depicts min and max of the IQR, which is colored.

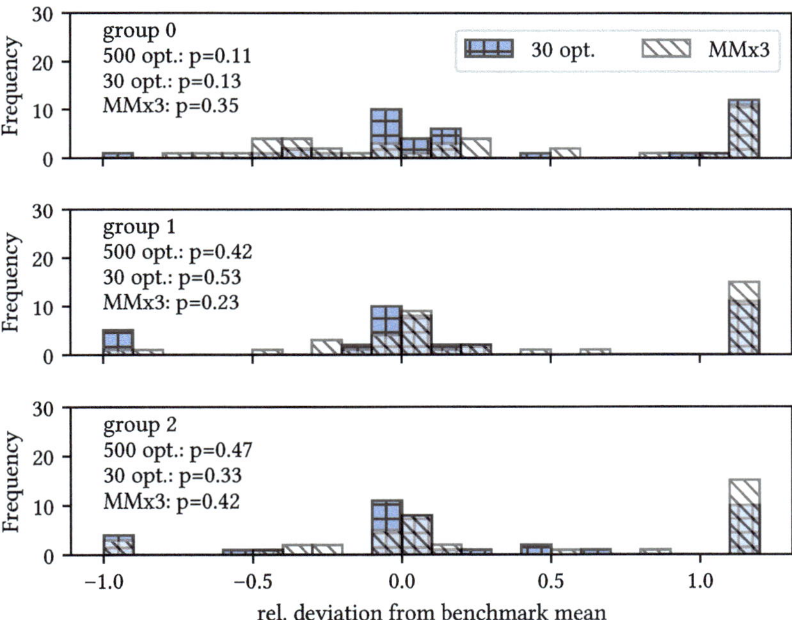

Figure C.27: Histogram comparing the reduced computational demand approaches regarding the relative difference from the mean of installed capacity of each technology to the benchmark inside of the clusters. Reduced computational demand at 30 optimizations.

Table C.2: Relative cost mean of input parameters, carbon emissions in million tons and cost of energy in € MWh^{-1} and CO_2 emissions in million tons by cluster in case of 80% emission reduction.

	all cases probability:	group 0 0.11	group 1 0.422	group 2 0.468
photovoltaic	0.53	0.35	0.69	0.43
onshore wind	0.55	0.59	0.55	0.55
offshore wind	0.27	0.28	0.27	0.27
battery, capacity	0.41	0.41	0.42	0.41
battery, power	0.41	0.40	0.41	0.41
PEM	0.24	0.24	0.24	0.24
alk. electrolysis	0.40	0.40	0.40	0.40
methanation	0.47	0.47	0.46	0.47
redox power	0.52	0.51	0.52	0.52
redox capacity	0.24	0.23	0.24	0.24
molten strorage	0.29	0.27	0.30	0.29
cost of energy	46.65	37.96	53.34	42.67
CO_2 emissions	58.52	19.28	128.82	4.47